Llamas & Alpacas

207
IVAN
5-30-03

Llamas & Alpacas

Small-Scale Camelid Herding

BY SUE WEAVER

An Imprint of BowTie Press®
A Division of BowTie, Inc.

Barbara Kimmel, *Editor in Chief*
Stephanie Staton, *Consulting Editor*
Lisa Barfield, *Book Design Concept*
Bill Jonas, *Book Design and Layout*
Indexed by Melody Englund

Reprint staff:
June Kikuchi, *Vice President Chief Content Officer*
Karen Julian, *Publishing Coordinator*
Jessica Jaensch, *Production Supervisor*
Tracy Burns, *Production Coordinator*
Cindy Kassebaum, *Cover design*

Additional photographs courtesy of Rhoda Peecher, Adrian Stewart, and Shutterstock.
Illustrations by Tom Kimball/BowTie, Inc.

Library of Congress Cataloging-in-Publication Data

Weaver, Sue.
Llamas & alpacas : small-scale camelid herding for pleasure and profit / Sue Weaver.
p. cm.—(Hobby farms)
Includes index.
ISBN 978-1-933958-57-6
1. Llamas. 2. Llama farms. 3. Alpaca farming. I. Title. II. Title: Llamas and alpacas.
SF401.L6W43 2009
636.2'966—dc22

2008046816

BowTie Press®
A Division of BowTie, Inc.
3 Burroughs
Irvine, California 92618

Printed and bound in China
16 15 14 13 12 11 10 2 3 4 5 6 7 8 9 10

This book is for Barbara Kimmel and Jarelle S. Stein—thank you, ladies, for your encouragement and endless patience—and for Deb Logan and Tina Cochran, whose love of lamas shines in "Advice from the Farm."

Table of Contents

INTRODUCTION

Why Lamas?

There has never been a better time than now to add llamas or fiber alpacas to your hobby farm menagerie. While top breeders still command impressive prices for the crème de la crème of the llama and alpaca world, it's becoming easier to buy correct, registered llamas and alpaca geldings at pet, performance, and fiber-owner prices.

Lamas (as llamas and alpacas are collectively called by those in the know) are fun to have around the farm. Their sweet, enchanting ways are sure to steal your heart. They cost little to feed and they're easy to handle, even by folk who have never kept livestock before. However, this is not to say they don't have specialized needs: feed, appropriate shelter, proper fences, and quality veterinary care head the list.

And that's what this book is about: the ins and outs of buying, understanding, caring for, and enjoying hobby farm llamas and alpacas. Read on, and consider the facts before deciding if lamas fit your lifestyle. If so, do your homework, prepare your farm, and then go lama shopping—and welcome winsome, wonderful llamas and alpacas to your farm and into your heart!

CHAPTER ONE

Meet the Lama

The Llama is a woolly sort of fleecy hairy goat,
With an indolent expression and an undulating throat
Like an unsuccessful literary man.
—Hilaire Belloc, *More Beasts for Worse Children* (London, 1897)

Llamas, alpacas, and their wild cousins, guanacos and vicuñas, are collectively known as South American camelids or simply lamas. Most people associate llamas and alpacas with South America's indigenous tribes, such as the ancient Incas, but few realize that the ancestors of these long-necked denizens of the Andes evolved in North America.

Lama History at a Glance

The oldest known protocamelid, a rabbit-sized, forest-dwelling creature known as *Protylopus*, appeared 40 to 50 million years ago during North America's Eocene era. The first true camelids evolved 12 to 24 million years ago. These included the genus *Paracamelus*, the ancestors of today's Old World camels. *Paracamelus* migrated north across the frozen Bering Strait about 3 million years ago and evolved into one-humped dromedaries and two-humped Bactrian camels. Some 2 million years ago, two more genera began migrating south through Central America into the South American Andes Mountains: *Paleolama* (which later became extinct) and *Lama*. *Lama* eventually evolved into two modern species: *Lama guanicoe* (the guanaco) and *Vicugna vicugna* (the vicuña).

Then, when referring to Pleistocene glacial epoch 10,000 to 12,000 years ago, a cataclysmic event occurred in North America that wiped out the remaining camelids there. Scores of other Ice Age mammals, such as the woolly mammoth and the saber-toothed tiger, also disappeared.

Archaeological evidence suggests that about 6,000 to 7,000 years ago, South American natives began domesticating wild camelids in the Altiplano (high plains) region of the central Andes Mountains, in areas now comprising southeast Peru, eastern Bolivia, and northern Chile and Argentina. The species that evolved there had to be tough and adaptable. A typical summer day in the Altiplano, which has an average altitude of 11,000 feet, may reach 70 degrees Fahrenheit, while nighttime temperatures may fall to 20 degrees or below. Between November and March, 90 to 95 percent of the year's 10 to 28 inches of rain falls; the rest of the year is very dry indeed.

Did You Know?

There were no wild llamas or alpacas. According to DNA studies conducted by American archaeozoologist Jane Wheeler and her colleagues, llamas are the domesticated descendants of wild guanacos, and alpacas are descended from wild vicuñas.

As bison were to North America's plains tribes, so lamas were to South

These llama-shaped bronze buttons follow the design of ancient effigies excavated at South American burial sites.

A Fortunate Foundation

Throughout prehistoric South America, llamas and alpacas were interred in human burials and buried en masse in important places. For instance, in the forecourt of the Chimú capital city of Chan Chan in Peru's Moche Valley (occupied from AD 1000 to 1400), priests interred hundreds of sacrificial llamas. Today, dried llama fetuses called *sullus* are buried under building foundations to bring good fortune, particularly in Bolivia, where an estimated 90 percent of families have at least one sullus buried beneath their homes. Construction workers refuse to work a job if there has not been a *cha'lla* (blessing ceremony) held before work begins and a sullus buried underground at the work site. Sullus can be purchased for a small fee from stands at La Paz's famous Witches' Market. Each comes blessed by a witch and is wrapped in *lana de llama*, a multicolored llama wool fabric.

America's early indigenous people—vital to survival. Llamas and alpacas supplied draft power, meat, fiber, grease, fertilizer, fuel, and leather. They were also precious for religious reasons, as evidenced by the many lama-shaped stone fetishes and conopas found at archaeological sites. Conopas, protective household figurines, had cavities in their backs that worshippers filled with offerings of intu (rendered fat from the chest of a llama) and coca leaves. So important did these symbols continue to be in native life that Spanish priests, seeking to convert the people by force in the seventeenth century, seized the conopas. Between 1617 and 1618, in the archbishopric of Lima alone, Spanish priests confiscated 3,418 conopas.

Naturally, lamas played an integral role in the lives of the great Incas, who flourished from the thirteenth to the mid-sixteenth century. By the end of the fifteenth century, the Incas controlled a 440,000-square-mile empire (covering much of present-day Ecuador, Peru, Bolivia, and Chile as well as parts of Colombia and Argentina) composed of more than 10 million people. Because sheep didn't come until much later, with the Spanish conquerors, everyone from the Sapa Inca (the divine ruler) to the littlest peasant child wore clothing woven of camelid fiber. The peasant had garments made of everyday llama fiber. The nobility dressed in garments of *campi*,

Llamas and alpacas are featured in an impressive array of South American craft items. This hand-carved stone llama is only 1 inch tall.

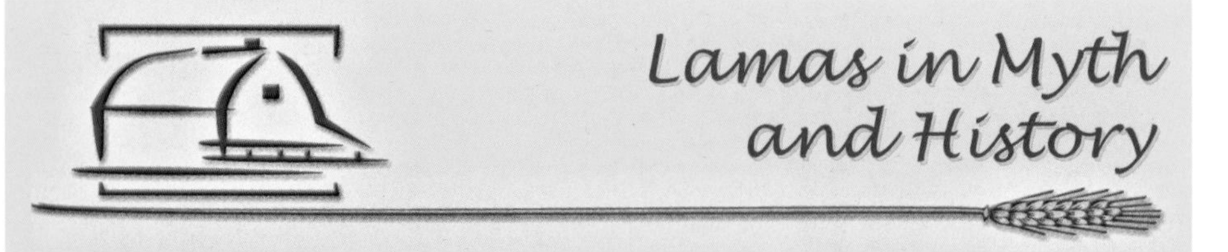

Lamas in Myth and History

Lamas were an important element in Incan religion. Black llamas, for instance, were considered rain bringers. In 1615, Spaniard Guaman Poma de Ayala wrote that at the beginning of the rainy season, the Incas tied black male llamas in the main plaza at Cuzco and left them without water so that they would cry out to Viracocha (the Incan creator god) for rain. Each day, a llama was sacrificed to the sun god, Inti, at sunrise, its head held toward the sun; the body was then burned in a special brazier.

The mythology of the Quechuas (a people of South America) tells of a celestial black llama called Yacana. In the middle of the night, Yacana drinks all the water out of the ocean. Should he ever fail to do so, the waters will drown the world. Yacana and Wiraqochan, the white alpaca, are responsible for nourishing the universe. Yacana appears in the night sky as a dark lane stretching from Scorpius to Centaurus in the Mayu (Milky Way). Another constellation in the Mayu is Uñallamacha, said to be a cria (a baby lama) attached to its dam (mother) by its umbilical cord.

Incan herders worshipped Urcuchillay, a multicolored llama who watched over their animals; his star is in the constellation modern astronomers call Lyra. According to Inge Bolin's 1998 book, *Rituals of Respect: The Secret of Survival in the High Peruvian Andes*, high-altitude herders still refer to llamas and alpacas as "our ancestors." When a llama becomes barren or its working days are over, it is slaughtered in an ancient ritual meant to speed the animal's spirit to Apu Illapu, god of thunder, who makes certain it is reborn to the same corral. The meat of the sacrificed llama is eaten, and the bones buried in the corral.

an ultrasoft fabric woven of vicuña fiber; no one else was allowed to wear it on pain of death. High-ranking officials wore garments crafted of *gami*, cloth woven of highest-quality alpaca fiber.

Although no exact figures exist, historians estimate the preconquest South American llama and alpaca population to have been as high as 50 million. Over the next hundred years, Spanish administrative documents indicate approximately a 90 percent reduction in numbers. Lamas were cleared in staggering numbers to make way for European species such as sheep, cattle, and goats. The people, too, perished in tragic numbers, from European diseases and overwork; thousands of South American people and African slaves died in the mines each year.

Fortunately, some natives and their llamas and alpacas fled to the high country of the Andes, where they survived through modern times. However, ancient lama husbandry practices were lost, and a great deal of crossbreeding between llamas and alpacas occurred after the fall of the Incan Empire.

In the mid- to late 1800s and early 1900s, private animal collectors and zoos both here and abroad began importing llamas. In the early 1900s, Californian William Randolph Hearst brought twelve llamas to his San Simeon estate, the largest North American importation up to that date. Then, during the 1980s, llamas became the exotic "critter du jour." Interest skyrocketed, and prices with it, until supply exceeded demand. Nowadays, high-end llamas still command impressive figures, but there are everyday llamas priced for the rest of us, too.

What's in a Name?

- The word *alpaca* is a derivative of the Spanish term *el paco*, which in turn comes from the Aymara word *allpacu*.
- In Spanish, *llama* can be roughly translated as "what is it called?" Legend claims that the Spaniards, having never seen llamas before, kept asking the Incas what they were called. ("¿Llama?") So the Incas thought that was the Spanish name for the animals.
- In Spanish-speaking countries, *llama* is pronounced YAH-ma instead of LAH-ma.
- In many countries, male alpacas are called machos and female alpacas, hembras.
- Names for lama hybrids include: *cama* (dromedary sire/guanaco or llama dam), *huarizo* (llama sire/alpaca dam), *misti* (alpaca sire/llama dam), *paco-vicuña* (vicuña sire/alpaca dam), *llamo-vicuña* (vicuña sire/llama dam), *llamo-guanaco* or *llanaco* (guanaco sire/llama dam), and *paco-guanaco* (alpaca sire/guanaco dam).

As bison were to Native Americans, lamas are to the Aymaran and Quechuan peoples.

Fiber Basics

Camelid fiber is hollow, so it is technically hair, not "wool," although it's commonly referred to as such. Fiber is measured in microns; a micron is 1/1,000 of a millimeter, or 1/25,000 of an inch. Alpaca fiber measures less than 20 microns (the standard grading system calls this "royal alpaca") to more than 35 microns (classification: "very coarse"), while llama undercoat generally grades from 20 to 40 microns. For comparison, vicuña runs 10 to 11.5 microns; guanaco, 14 to 18 microns; Angora rabbit fiber, 12 to 16 microns; and fine Merino sheep wool, 18 to 22 microns. The lower the count, the finer the fiber. Yarn containing more than 5 percent fiber measuring 22 microns or greater is generally too coarse and itchy to wear next to human skin.

Alpacas

Fiberwise, there are two types of alpacas: huacaya and suri. *Huacaya* (h'wha-k'EYE-ya) alpacas have crimped, plush fleece and cute, teddy bear faces. Because they're more common than suris, huacaya alpacas generally cost less to buy. If you want fiber to knit or crochet with, huacaya fiber is more economical and warmer than sheep's wool.

Alpacas coat types: huacayas (left) and suris (right)

Suri (SIR-ee) alpacas' fiber falls in long, lustrous locks that separate into individual ringlets. The relative rarity of suris accords them extra value. Suri fiber is used for weaving "worsted" items such as the fabric used to craft fine suits and overcoats, where drape is important and elasticity isn't an asset. It can also be knitted into sumptuous lace with a silklike sheen.

Llamas

Suri llama fiber

Some types of llamas are double-coated, and others aren't. The fiber of double-coated llamas is composed of up to 20 percent guard hair, which must be removed before the undercoat is processed into yarn; otherwise, the resulting yarn is bristly and itchy. Dehairing sheared or clipped fiber is usually done by hand. Because it's a slow, painstaking process, spinners should avoid llamas with overly abundant guard hair.

Suri llama fiber resembles suri alpaca fiber and is used in the same manner. The Suri Llama Association and Registry actively promotes these rare and regal beasts. Apart from suris, llamas fall into one of two basic categories: classic or woolly-coated. Many people simply call them short-, medium-, or long-wooled llamas; others use the following terminology.

Classic llamas, also called *ccara* (CAR-uh) or *ccara sullo* (CAR-uh SOO-yoh) llamas, have relatively short, double coats. The amount and length of a ccara llama's guard hair varies greatly from individual to individual but usually accounts for 15 percent or more of its overall fleece. Ccaras have soft, semicrimpy undercoats topped by coarser guard hairs and shorter hair on their heads and legs, especially below the knees. Ccaras shed their undercoats, so they needn't be sheared or clipped. It's easy to harvest ccara fiber by grooming its wearer and removing shed hair from the grooming tools. Ccara llamas yield about 1 to 3 pounds of fiber per year. Ccara coats don't pick up debris the way the coats of longer-wooled llamas do, and they're easy to keep clean. This makes them the preferred type of llama for packing (a job they were historically developed to do) and public relations work. They are sometimes referred to as "zoo llamas" because most early imports were ccara llamas.

Suri llamas are rare and in high demand.

Most llamas have guard hair to some degree.

Medium-wooled *curaca* (cur-AH-cah) llamas resemble their ccara kin but have less guard hair (3 to 15 percent on average) and longer wool on their bodies, necks, and legs. Like ccaras, curacas have hair instead of wool below their knees and hocks. Curaca llamas partially shed, but most require clipping or shearing every few years.

Tapada (tah-PA-dah) llamas are woolly llamas with medium to long coats that are dense, sometimes silky, sometimes crimpy, often loosely wavy coats. They also have wool (not hair) on their heads and below their knees. They have less than 1 percent guard hair and are considered single-coated. Longer-coated tapadas are frequently confused with *lanuda* (la-NOO-dah) llamas. Lanudas are single-coated, silky-woolly, long-coated llamas with wool-fringed ears and tails, well-wooled faces, and abundant fiber all the way down their legs to their feet. Tapada and lanuda llamas don't shed; they require full body shearing every year or so and need considerable grooming in between. They yield the most (and usually the best) fiber.

This llama's guard hair is longer than its undercoat.

Alpacas are another story. Although groups were imported to the United States in 1821 and 1857, these failed to thrive. Then in the 1980s, alpaca fever took root here. Large-scale alpaca importations into the United States occurred between 1984 and 1998. In 1998, the Alpaca Registry, Inc., (ARI) closed its herd book to imported alpacas. Since then, all alpacas recorded in the official herd book must have two ARI-registered parents and be born and bred in the United States. At this writing, demand far exceeds supply, and top-quality females and stud males command astounding prices. However, the market for males that are not up to stud male standards is becoming saturated, making fiber geldings (castrated males) a viable choice as pets and for hobby farm fiber producers.

This handsome suri alpaca's topknot hides his face.

Did You Know?

- The Inca were known to raise pure white, black, and brown alpacas. The weaving of alpaca fiber textiles dates to at least 500 BC.
- Authorities believe llamas and alpacas were developed from guanacos and vicuñas long before their Old World cousins, dromedaries and Bactrian camels, were domesticated (around 3000 BC in South Arabia and Iran, respectively).

Lamas 101

South American camelids belong to the taxonomic order Artiodactyla (even-toed ungulates), suborder Tylopoda (pad-footed ungulates), and family Camelidae. There are four species: the guanaco (*Lama guanicoe*), vicuña (*Vicugna vicugna*), llama (*Lama glama*), and alpaca (formerly *Lama pacos*, now reclassified as *Vicugna pacos*).

The Wild Ones

Wild vicuñas and guanacos still roam the Andes, as they have for thousands upon thousands of years. Vicuñas are considered a threatened species. Because they can't legally be exported from South America, you cannot buy vicuñas to stock your hobby farm. However, a limited number of domestic guanacos are bred in North America, so they are an option should you want some.

Vicuñas

Vicuñas are the wild ancestors of alpacas. They are the smallest and daintiest of the South American camelids, standing about 3 feet tall at their shoulders and weighing 100 to 120 pounds. Vicuñas are exceedingly shy and can flee at speeds of up to 35 miles per hour, aided by a heart almost 50 percent larger than the average weight for mammals of similar size. They are svelte and streamlined, having long, skinny legs and necks and small wedge-shaped heads.

Wild vicuñas roam the Andes mountain range of Peru, Bolivia, Argentina, and Chile, at elevations of 12,000 to 19,500 feet. They are Peru's national heritage species; hunting them is strictly forbidden, but natives, who were granted ownership of the nation's vicuñas in 1987, are allowed to capture, shear, and release wild vicuñas every two years.

Typical coat colors range from yellowish ochre to reddish brown, always with dirty white underparts and a white mane of 8-to-12-inch silky hairs at the base of the neck. Each vicuña produces about 12 to 16 ounces of 10–12 micron fiber with a staple length of 1–1.5 inches per semiannual shearing. Because vicuña is the ultimate luxury fiber, a vicuña scarf may sell for $800 and a finely tailored vicuña suit for up to $20,000.

The Inca never killed vicuñas because they were the sole property of the Sapa Inca. They trapped vicuñas in *los chacos*, a massive ceremony held every three to five years in which hundreds of people formed a human chain to herd the animals into temporary corrals where they were shorn. Vicuñas are still gathered in the same manner.

In preconquest South America, all vicuñas belonged to the Sapa Inca. Even today, the Peruvian government prohibits exportation of live vicuñas.

Lamas by the Numbers

- Peru is the world's largest producer of alpaca fiber at 4,000 tons per year and vicuña fiber at 3 tons per year, while Bolivia is the largest producer of llama fiber at 600 tons per year.
- According to an *International Camelid Quarterly* survey conducted in 2004, 66.9 percent of alpaca and llama farms keep fewer than 25 lamas, and 60 percent are only 1 to 20 acres in size.
- There are roughly 3 million alpacas worldwide; 98 percent of the world's alpacas are in Peru, Bolivia, and Chile. Of these, 90 percent are huacayas; the rest are suris.
- Only 3 percent of America's registered llamas are imported from South America.

Prior to the Spanish conquest, the Sapa Inca owned an estimated 2 million wild vicuñas. However, the Spaniards slaughtered more than 80,000 vicuñas annually, partially to clear the way to graze sheep and cattle and partially for their meat, fiber, and hides. In 1825, when Peru gained its independence from Spain, vicuñas became a national symbol. However, by 1974, when an inventory was taken, only 6,000 wild vicuñas remained. Now the Peruvian vicuña population is rebounding, growing 8 percent a year, and the species is no longer in danger of extinction. There are 149,000 vicuñas in Peru and 15,000 in Bolivia; experts say that if recovery proceeds at the present rate, by 2021 there should be 1 million vicuñas in Peru alone.

Guanacos, native to South America, are bred in North America, too. If you want some, they're out there!

Llamas can easily be trained to pull carts. This one is working at the London Zoo in days gone by.

Although you can't buy purebred vicuñas in North America, enterprising lama entrepreneurs are breeding paco-vicuñas (alpacas with vicuña DNA and physical characteristics) and Pacuñas (true vicuña and alpaca hybrids) in an effort to produce vicuña-quality fiber in greater volume than true vicuñas grow each year. See the Resources section for the organizations that register and promote these interesting creatures.

Guanacos

Guanacos are the wild ancestors of llamas. Like llamas, they have a double coat consisting of coarse guard hair covering a soft undercoat even finer than alpaca fiber. Guanacos resemble vicuñas in many ways, being swift, wary, and slender, with long legs and necks, small heads, and pointed ears. But they are taller and somewhat heavier, standing 3 to 4 feet at their shoulders and weighing in at 220 to 275 pounds. Guanaco colors range from light honey brown to dark cinnamon, always with dirty white underpinnings.

Prior to the Spanish conquest, an estimated 30 to 35 million guanacos lived in South America; the current wild population numbers roughly 500,000. They're found from sea level at Tierra del Fuego at the southernmost tip of Argentina to the high Andes of northern Chile and Peru.

Guanaco undercoat averages 14 to 18 microns in diameter with a staple length of 1.5 to 2.2 inches; a typical guanaco produces 350 to 600 grams of fiber per shearing.

Guanaco fiber is highly prized by handspinners and, with considerable early handling, guanacos make nice

Llamas and Alpacas at a Glance

Typical adult height measured at the shoulder
Llamas: 4–4.5 feet
Alpacas: 3 feet

Typical adult weight
Llamas: 250–500 pounds
Alpacas: 120–225 pounds

Colors
Llamas: white to black and many shades of brown and gray in solids, spots, and a wide variety of patterns.
Alpacas: in North America there are 22 recognized colors ranging from white to red to black and many shades in between; 52 colors are recognized in South America.

Primary reason for development
Llamas: as pack animals
Alpacas: for fiber

Annual growth rate of fiber
Llamas: varies widely
Alpacas: 5–10 inches

Weight of a typical adult fleece
Llamas: varies widely
Alpacas: 3–8 pounds

Fiber measurement in microns
Llama: 20–40 microns
Alpaca: 15–28 microns

Life span
15–25 years

Length of a typical gestation
Llamas: 350 days
Alpacas: 345 days

Weight at birth
Llamas: 18–35 pounds
Alpacas: 12–18 pounds

Shape of ears

Llamas: long and banana-shaped

Alpacas: shorter and spear-shaped

Shape of back in profile
Llamas: straight
Alpacas: somewhat rounded

Rectal temperature
99.5–102 degrees Fahrenheit

Heart rate
60–90 beats per minute

Respiration
15–30 breaths per minute

pets. To locate breeders do an online search using the keywords *guanacos for sale*. The International Lama Registry (see Resources) maintains a herd book for registered guanacos.

Everyday Lamas

Few of us will ever own a guanaco or see vicuñas running wild. However, the wild ones gave us the lamas we know and love: regal llamas and cute woolly alpacas.

Llamas

According to *International Camelid Quarterly* statistics gathered in December 2004, there were 170,000 llamas and 635 guanacos and lama hybrids registered in International Lama Registry herd books at that time; an additional 5,768 llamas were registered in 2005. In addition, the International Lama Registry estimates that at least 10 percent of the American llama population is unregistered. Clearly, llamas are popular indeed!

Americans use their llamas for a wide range of activities, ranging from serving as elegant pasture pets and livestock guardians to 4-H activities, cart driving, packing, and public relations work such as visiting shut-ins and children's hospitals. They can be shown in a wide array of halter events (also called conformation or in-hand classes) as well as a long list of performance classes, including costume, packing, and agility. Most llamas produce lovely fiber as well.

Lamas—like this cute alpaca cria—are curious to a fault.

Alpacas

The same *International Camelid Quarterly* study determined that the ARI had registered 67,608 alpacas by the close of 2004. That, however, is only the tip of the iceberg. Alpacas have become so popular in the past few years that the organization registered 19,755 new alpacas in 2007 alone, for a grand total of 136,075 alpacas registered through December 2007.

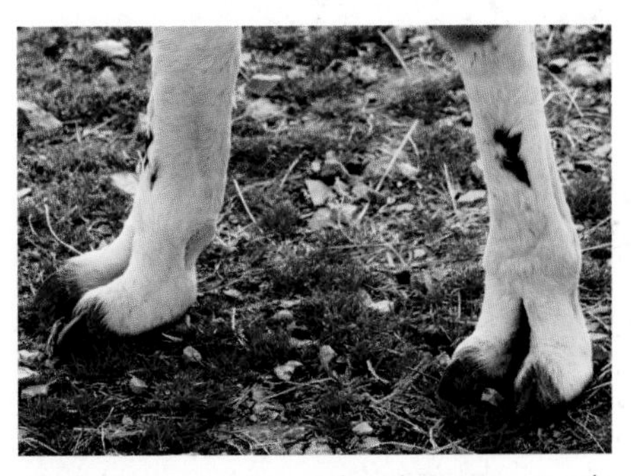

Lamas' toes are topped with strong, sturdy toenails.

Although alpacas are largely fiber producers, they can do most of the things their larger cousins can do. They especially shine as public relations lamas (what shut-in isn't cheered by an alpaca's endearing face?). And alpacas have their own array of halter and performance classes at llama and alpaca shows; showing is a great way to further enjoy your fiber geldings.

Basic Lama Physiology

Lamas of all kinds share features that make the South American camelids truly unique. They are all ruminants (they chew their cud), but they have three-compartmented stomachs instead of the usual four found in sheep, goats, cattle, and deer. Their two-toed feet have broad, doglike leathery pads on the bottom of each toe, with down-curved nails in front and a scent gland between the toes. The oblong bare patches on the side of each rear leg are metatarsal scent glands associated with the production of alarm pheromones.

All lamas have a hard upper dental palate and no upper teeth in front; upper and lower molars in back; and a split, prehensile upper lip for grasping forage in

Young males, like these handsome youngsters at Klein Himmel Llamas, enjoy engaging in play fighting games.

TERMS USED TO DESIGNATE SUPERFICIAL AREAS OF THE BODY OF AN ALPACA

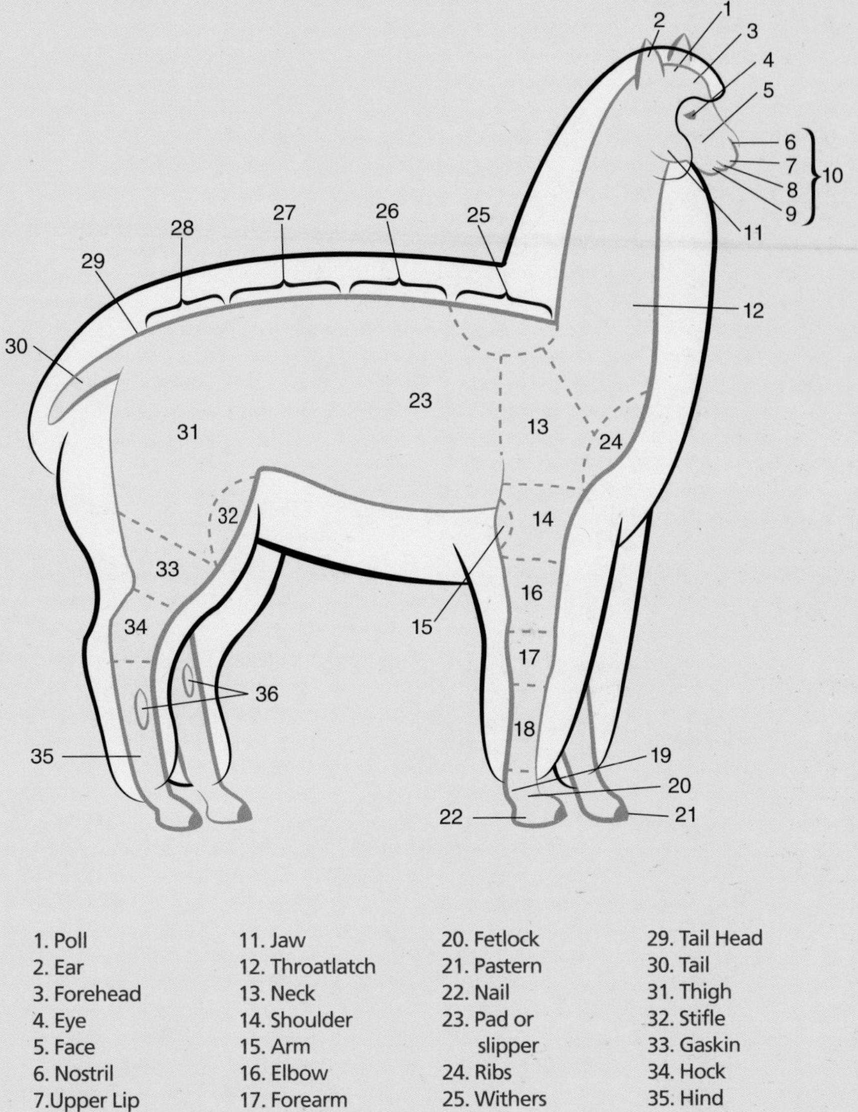

1. Poll
2. Ear
3. Forehead
4. Eye
5. Face
6. Nostril
7. Upper Lip
8. Mouth
9. Lower Lip
10. Muzzle
11. Jaw
12. Throatlatch
13. Neck
14. Shoulder
15. Arm
16. Elbow
17. Forearm
18. Knee
19. Cannon or shank
20. Fetlock
21. Pastern
22. Nail
23. Pad or slipper
24. Ribs
25. Withers
26. Back
27. Loin
28. Croup
29. Tail Head
30. Tail
31. Thigh
32. Stifle
33. Gaskin
34. Hock
35. Hind Cannon
36. Scent Gland

Advice from the Farm

Why Lamas?

The experts offer a few good reasons for welcoming lamas to the farm:

Inexpensive Perfection

"Right from the beginning, it was clear that alpacas were the perfect livestock for us. They are cute and soft, quiet, don't take up much room, and best of all, they all poop in the same place! We heard they were only for the rich, but we had to see them in person anyway. Once we met our first alpaca, there was no question; we wanted some. But was it possible to start our alpaca farm without spending our entire life savings? Yes! Our solution: we keep only herd sires and provide mobile breeding services at our customers' farms. It's possible to afford great alpacas if you think outside the box."

—Tina Cochran

Wicked Smart

"Llamas have a certain grace and gentle elegance that attracts me. I am drawn to the fact that I must earn their trust in order to interact with them. They are 'wicked smart,' responsive, and so very kind and gentle. To be sitting amongst them is akin to being surrounded by unicorns."

—Deb Logan

Elegant Guardian

"I adopted Alex, my llama, to guard my sheep. We have close neighbors, so we can't keep a livestock guardian dog that barks all night, and my husband is afraid of donkeys (silly but true), so a llama seemed to be our best choice. Alex is elegant and quiet, he's affectionate without being pushy, and he loves my sheep. You should see how gentle he is with newborn lambs!"

—Jan Johnson

Yummy Fiber

"As a longtime handspinner, I thought it would be fun to keep llamas to grow my own fiber. Then I saw an ad for unregistered male alpacas at surprisingly low prices. I called and made an appointment, and when I saw the alpacas I was blown away! They were just babies but so cute and so soft. I bought two and named them Cash and Cary. Their fiber is yummy, and they are wonderful pets. I'm going to have Cash and Cary gelded when they're old enough and buy two more when this year's crias are weaned. I'm going to process their fiber and sell it on eBay—but I get to keep what I want for me!"

—Mary Collins

Lamas rest in this kushed position, with their legs tucked under their bodies.

unison with the lower incisors. Mature males grow long, curved fanglike "fighting teeth."

Lamas rarely touch one another and, unless habituated to it, prefer not to have humans touch them, either. Because when males fight, they try to pull their adversary's legs out from beneath him, untrained lamas vigorously resist having their legs picked up. Although lamas sometimes lie on their sides or backs, they usually rest in the kush (or cush) position with their legs tucked under their bodies.

Mating occurs in this position, with the male orgling (emitting a mating-specific vocalization) and grasping the female's sides with his front legs. Males are dribble ejaculators, so a typical mating takes twenty minutes or longer. Female lamas are induced ovulators, meaning they have no heat or estrus cycle. Instead, the act of breeding causes the female to ovulate about a day and a half after mating.

Although they generally produce single young (called crias), female lamas, like cows, have four teats. Crias are nearly always born during daylight hours and from a standing or squatting position rather than lying down. Because lamas have attached tongues, they don't lick their newborns in the manner of many other species. Twinning is a rare occurrence.

Lamas communicate in body language and through a variety of vocalizations ranging from a gentle hum to loud shrieks. And they spit—generally only at one another but occasionally at humans, too.

Lamas instinctively eliminate on community dung piles instead of randomly wherever they are. When one eliminates, usually all of the lamas in a group will stand in line to do so as well.

Ultra-tidy animals, lamas deposit their dry, formed droppings (also known as lama beans) on community dung piles.

CHAPTER TWO

Buying a Lama and Bringing It Home

Llamas and alpacas are fun to own and relatively easy to care for. However, when you buy or adopt one, you're taking on a good deal of responsibility. As with all animals, they will need to be tended to every day, whether you feel like it or not. Your lamas will depend on you for their very lives. Think it over carefully before you commit. And while llamas or alpacas fare well on smaller rural properties, make certain you can legally keep them; check into local zoning laws before you bring some home.

If after doing your research and considering all the factors, you're certain you're ready to buy your lamas and know which kind you want (see chapter one), the next step is for you to find the right seller. And once you've made your purchase, you'll need to find a safe way to get your new lama home and make sure that you are prepared to receive it properly, including lining up a lama-savvy veterinarian beforehand.

Finding a Reputable Seller

No matter what you're looking for—show or fancy fiber alpacas, pack llamas, or a nice guard llama to protect your sheep or goats—it's important to locate reputable sellers of healthy stock. To do so, consider these approaches.

Visit the Web sites for the International Lama Registry (if you're looking for llamas, guanacos, vicuñas, or crosses thereof), the Suri Llama Association and Registry (suri llamas), the American Miniature Llama Association (miniature llamas), or the Alpaca Owners and Breeders Association (alpacas), and peruse member-breeder directories there. Phone or e-mail organizations for additional information.

New Owner's Checklist

So you've thought it over and are sure you want lamas in your life. But are you ready to bring some home? You will need:

- Safe shelters, bedding, feed, proper fencing, feeders, water containers, and a clean, consistent water source. Read about these items in chapters four and five.
- Halters, leads, toenail trimmers, and shearing necessities.
- A first aid kit.
- Appropriate tack, if you plan to pack or drive your llamas.
- Safe birthing quarters and a well-stocked birthing kit, if there are crias in your future.
- And most important: phone numbers of at least three veterinarians who treat llamas and alpacas (not all do). If the vet you want isn't available, you should know of at least two others you could call when you desperately need one.

Take a look at online breeder directories. To find them, type *llama directory* or *alpaca directory* in the search box at your favorite search engine. Visit breeders' Web sites. Type *llamas* or *alpacas* and *sale* into your search engine's search box. You could also qualify it by state (*llamas sale Montana*). If breeders' Web sites don't offer the llama or alpaca you're searching for, e-mail sellers to ask whether they have additional stock for sale. If they don't have what you want, they may know someone who does.

There are eleven great llama and alpaca periodicals listed in the Resources section of this book; they all run display ads, directories, and classifieds. Subscribe to your favorites or pick them up at newsstands and farm stores such as Tractor Supply Company (TSC).

Join llama- and alpaca-related e-mail groups where subscribers post what they have to sell. It's a great way to find lamas and lama-related supplies while making new (lama-knowledgeable) friends.

Take in a show, a seminar, or an expo. Visit information booths, and chat with exhibitors between classes; make friends, and meet lamas! State and regional lama associations sanction shows and host seminars and expos; e-mail or call organizations for specifics.

Watch for "llama for sale" notices on bulletin boards at feed stores and veterinarian practices, or pin up "llama wanted" notices of your own. Monitor local classified ads. Talk to vets and county extension agents in your buying area; they're sure to know who raises llamas and alpacas in your neck of the woods.

If you want to breed llamas or alpacas, start with stock from other successful breeders. Study show results, read pertinent material in magazines and online, and talk with other breeders producing the sort of stock you'd like to own. Who is raising the sort of llamas or alpacas you want to own? If you can, go straight to the source.

If you can buy the stock you need in your locale, do so. Healthy animals acclimated to your region and spared the stress of long-distance travel tend to

Visit a breeding farm before purchasing to ensure your lama comes from a clean environment.

arrive and remain healthier than those trucked in from distant sources. However, if you plan to breed high-end show and breeding stock, you may not have a lot of local options. You'll probably have to buy from afar, perhaps through a production sale or via the Internet.

Buying from Breeders Near and Far

Once you've narrowed the field to a handful of people selling your type of llamas or alpacas, contact them, and if possible, arrange to visit their farms.

Be courteous and arrive at the designated time. Look around. Farms don't need to be showplaces, but they shouldn't be trash dumps, either. Are all types of livestock, not just the lamas you came to see, housed in safe, reasonably clean

Did You Know?

According to a lama industry survey conducted in September of 2004 by *Camelid Quarterly* magazine (based on questionnaires mailed to 45 percent of North America's camelid industry participants):

- 70.5 percent of U.S. respondents raise llamas, 31.5 percent raise alpacas; 62.5 percent of Canadian respondents raise alpacas, 37.5 percent raise llamas.
- 25 percent of the participants have raised lamas for more than ten years.
- There are 1–25 lamas on 66.9 percent of U.S. farms; only 1.7 percent own 101–250 lamas.
- 60 percent of U.S. farms and 45.6 percent of Canadian farms are of 1 to 20 acres in size.

Advice from the Farm

What to Do Before You Buy

The experts offer advice to first-time lama buyers.

Research, Join, and Consider

"Do as much research as you possibly can before you buy or adopt. An Internet search will turn up a plethora of sites, and each of those will have links to many more. Some outdated information is still being disseminated, but after you run through a number of sites, you will be able to tell which info is up to date and which isn't.

"Join as many llama and alpaca e-mail groups as you can. For alpacas, the primary list to join is *alpacasite* at YahooGroups; it hosts more than 3,500 members and is a very active list. Llama lists are plentiful, although some are more active than others. When you find a group, look at the number of messages by month and choose the ones that are active. Once you've joined a list, go back through the archives and move forward through the messages—they cover anything and everything you ever could think to ask! You can sort by date or by subject.

"Consider your environment—don't purchase dark-colored, heavy-wooled animals if you live in Florida (unless, of course, you have a pond they can stand in and you don't mind that it will rot their fiber).

"Learn how to body score a llama before you go shopping, and put your hands on the animals to ensure they have had proper nutrition. All that wool can hide a *lot* of problems."

—Deb Logan

Listen to Yourself

"Before investing in alpacas, make sure you've done your homework and have an idea of what you want from them. Read all you can, visit breeders, go to fiber shows, and, most importantly, expose yourself to alpacas as often as you can.

"Then, don't buy alpacas the first time you see one. Each breeder has different ideas of what makes a certain alpaca 'great.' Listen to yourself, and figure out what you enjoy about alpacas instead of what is the latest thing."

—Tina Cochran

Choose Wisely

"When you're ready for llamas or alpacas, remember that camelids are herd animals and will slowly decline or can become aggressive if they're kept alone. Camelids raised in isolation have no idea how to act appropriately for their species and will be shunned (and sometimes harmed) if later introduced to a herd.

"Unless you intend to breed, it's easiest to pick a gender and stick with it. Even some geldings will continue to 'sneak breed,' which can lead to infections and scarring in the females because of the intrusive nature of the breeding process. (A camelid's penis actually enters the uterus itself and spins.)

"Never purchase an 'over-friendly' male. He may morph into a dangerous animal when he's grown. The vast majority of aggressive males surrendered to rescue groups were bottle raised and/or overhandled as youngsters and no longer respect people. They view them as contenders for their space, their females, and often even their food."

—*Deb Logan*

Remember: It's a Commitment

"Keep in mind that alpacas are easy-care livestock, but they are livestock. They require daily care and yearly shearing; they are not a 'put out in the field and come back in six months' type of animal. You will be shoveling 'beans,' lifting bales of hay, and getting dirty. Keeping livestock is a 24–7 commitment. If you only want to have alpaca fiber to use, buy the fiber, not the animal."

—*Tina Cochran*

facilities? Are they in good flesh, neither skinny nor roly-poly fat? In large herds, you're apt to spot individuals that are skinnier or fatter than the norm, but the majority should be in just-right condition.

Ask about the seller's vaccination and worming philosophies. Which vaccines and wormers does he use and why? How often does he vaccinate and worm his llamas or alpacas?

If you like what you see, ask to examine the llama's or the alpaca's registration papers as well as its health, vaccination, worming, and production records. Specifically inquire about a female's birthing habits. Has she had any problems giving birth? Is she a good mom? Have her offspring excelled at the very things you're breeding for?

Find out whether the seller is willing to work with you after the purchase, should questions or problems arise. Get any guarantees in writing.

And trust your intuition. If a seller seems evasive or makes you uneasy for any reason, thank him for his time and go on down the road. There are too many honest llama and alpaca people out there to deal with someone you don't quite trust.

Production and Dispersal Sales

Production and dispersal sales are fine venues for buying quality animals at fair market prices. The best are publicized well in advance and offer printed catalogs that highlight sale animals' pedigrees and production records. To

Take a look at the surroundings when you visit prospective sellers. The entrance to Klein Himmel Llamas' main barn is neat and clean. It says "we care about our llamas here!"

find out about these sales, peruse lama magazines and visit the Web sites of llama and alpaca organizations. Production sales are llama and alpaca world social events and can be the perfect places to meet people and purchase quality lamas.

Payment in full is expected on sale day. Animals sell with registration papers, health certificates, and any other documentation needed for interstate shipment. Guarantees, if any, are stated in the sale catalog.

Major sales make provisions for absent bidding (usually by phone, fax, or e-mail), but it's better to be there in person. Arrive before the sale starts, and do hands-on inspections of all of the animals you think you might bid on. Study the catalog to see which other lots were consigned by their owners and give those lamas a once-over, too. Another sensible ploy: mark your catalog, designating which animals you plan to bid on, and make a notation of your absolute top bid.

Internet Buying

There are lots of good reasons to shop the Internet when you're looking for livestock. You can shop for llamas or alpacas anywhere in the world, at any time of day, seven days a week, without leaving home, and you can select from a vast pool of animals and breeders. You can research interesting animals and sellers before you deal, thus saving a great deal of time and money on farm visits. However, be especially careful to deal with reputable sellers. Request buyers' references and check them out.

Ask for video footage of animals that interest you. If it isn't available, ask to see additional photos taken from many different angles. Examine these materials very carefully and address any issues before you buy. Insist on a written guarantee, and negotiate its terms before you commit.

Be clear about how you will get your purchases home before you make a deposit. Who pays for interstate health papers and the tests they entail? How long will the seller hold the animals once payment is made and you're working to line up transportation? Who foots vet bills incurred during that wait? What happens if an animal should die? Get it in writing; don't leave anything to chance.

Sale Barns

The first rule of llama and alpaca buying is to buy from individuals or at well-run dispersal and production sales, never from neighborhood sale barns; they're dumping grounds for sick animals and culls.

If you buy at sale barn auctions, you won't know if the animal you get has been vaccinated or what's going on in the herd of origin. You won't know whether a female is pregnant and, if so, by what male. A male may be infertile—or so dangerous that his owner is willing to see the last of him at any price. And animals that weren't sick or exposed to disease before they're sold through an auction will be by the time it's over.

If you attend such sales, even just to look, you're going to be tempted to buy. If you do it, remember: never, *ever* turn newly purchased sale barn lamas out with the rest of your herd. Always, without exception, quarantine them for at least three weeks.

Registration Papers

Let me repeat: carefully check the registration papers of any llama or alpaca you're buying to make certain you're getting what you're paying for.

The animal must be recorded in the seller's name. The last recorded owner is the only person who can transfer the animal to you. In some cases, the seller may have a transfer form made out in his name but has never gotten around to having the papers

Klein Himmel Llamas' animals are neat, clean, and happy. This is the sort of establishment to buy from.

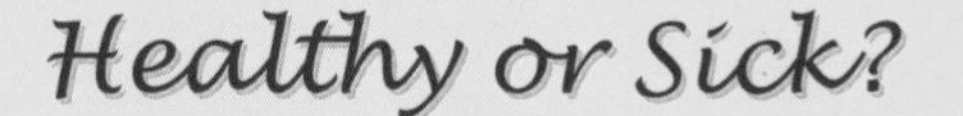

Healthy or Sick?

No one wants to end up buying a lama that is already sick. Here are some signs to look for when inspecting prospective purchases.

Healthy Lamas	*Sick Lamas*
Are alert and curious.	Are dull and disinterested in their surroundings. May isolate themselves from the rest of the herd.
Stand or graze in normal postures.	Stand with head drooped, ears back or lopped, with a slightly rounded back (keep in mind that the backs of alpacas are naturally more rounded than those of llamas). They may grind their teeth, groan, and look or kick at their bellies.
Hum only at appropriate times.	Continually hum; sometimes in a low, depressed-sounding manner, sometimes rather stridently.
Have bright, clear eyes.	Have dull, sunken, or depressed-looking eyes.
Have dry, cool noses. (A trace of clear nasal discharge isn't cause for concern.)	May have fresh or crusty opaque discharge residue in the corners of their eyes. May have thick, opaque, creamy white, yellow, or greenish nasal discharge.
Have regular and unlabored breathing.	May wheeze, cough, or breathe heavily and/or erratically; breathe with their mouths open; or drool.

Healthy Lamas	*Sick Lamas*
Have no unusual lumps on their jaws, throats, umbilical area, or anyplace else on their bodies.	May have a significant lump on one jaw (indicating an abscessed tooth; this can be quite serious) or lumps—sometimes oozing pus—elsewhere on their bodies.
Have healthy-looking fleece and pliable, vermin- and eruption-free skin.	May experience wool loss; skin may show evidence of external parasites or skin disease.
Move freely and easily.	Move slowly, unevenly, or with a limp; may be significantly uncoordinated, especially in the hindquarters.
Willingly rise from kushed position when urged to do so.	Often refuse to rise despite vigorous encouragement.
Seem to be of average height and weight for their age.	May be thin or emaciated; conversely, may be grossly overweight, which can lead to heat stress and fertility problems.
Have healthy appetites; ruminate (chew cud) after eating.	Won't eat (usually ill); don't ruminate (chew cud) (very ill indeed).
Have firm, slightly elongated, berry-like droppings; tail and surroundings areas are clean.	May have scours (diarrhea); tails, tail area, and hair on hind legs may be matted with fresh or dried diarrhea.
Have normal temperatures (99.5–102 degrees Fahrenheit for an adult lama; slightly higher values are acceptable on hot summer days and immediately after exercise).	Run high or low temperatures; subnormal temperatures are generally more worrisome than fevers are.

updated. Some (but not all) registries allow this type of seller to issue a second form transferring his impending ownership to you; if you (and your registry) accept this arrangement, be prepared to pay double transfer fees.

If you're buying a bred female, you'll need a breeder's certificate signed by the owners of both the male and the female in order to register her cria. Depending on the registry you do business with, the service memo may be a separate document or part of the transfer slip. Contact your registry for particulars before you go shopping.

Before handing over your check, get all applicable guarantees and sales conditions in writing. Do this every time, even when dealing with friends. People misunderstand, and people forget. Hammer out the details, and write them down!

Quarantine

When you bring a llama or an alpaca home from *any* sale, plan to quarantine it away from your existing herd. House the new animal in an easy to sanitize area at least 50 feet from any other llamas or alpacas, but where it can see other lamas at a distance. Worm the lama, vaccinate it, trim its toenails if needed, but keep it isolated for at least three weeks. Don't forget to sanitize the conveyance you hauled the lama home in. During that time, feed and care for your other animals first, so you can scrub up after handling the new addition. Never go directly from the quarantine quarters to the rest of your herd. If you can prevent it, don't allow dogs, cats, poultry, or other livestock to travel between one group and the other. When the new lama's time in quarantine is up, sanitize the isolation area and any equipment you've used.

Methods for Transporting Lamas

Llamas and alpacas are both the easiest and the hardest of livestock to transport. A well-socialized, well-trained pack llama or a pair of seasoned public relations alpacas will hop in the back of your SUV and kush, and down the road you'll go with not so much as a lama "bean" to sweep out at trip's end. (Llamas and alpacas eliminate only on established potty piles.) However, hauling stressed, unhandled llamas or alpacas under adverse weather conditions, even in the best of lama conveyances, can be a nightmare. Here are some things to consider.

Working with Livestock Transporters

In many cases, transporting your own llamas or alpacas isn't cost-effective. If you're relocating to a new home a thousand miles away and taking your camelid friends with you, or if you've purchased your dream female and her cria from a breeder in a distant state, you may want to pay a livestock transporter to bring your lamas to you.

To find these professionals, try a search for *livestock transport* at your favorite search engine. You can also check out lama-related e-mail lists for postings. Many professionals as well as weekend transporters (for example, owners who are transporting their own animals and are willing to carry a few extras for pay) post their trips to these lists—another great reason to subscribe to one or more.

Obtain estimates from at least three or four companies (prices vary dramatically from hauler to hauler), and ask for references, then take time to check them out. Unless you contract for the whole load, routes are rarely direct, and your llamas or alpacas may be in transit for days (occasionally, weeks). Because lamas are easily stressed, it's important to choose a transporter who knows them and goes the extra mile to keep them safe and well.

Find out what sorts of tests and health papers are necessary for your lamas' interstate shipment, and have the paperwork in hand when the hauler arrives to pick your lamas up. (Responsible haulers will refuse to haul your animals without these documents.)

Questions for Transporters

When comparing rides, ask the transporters:

- What sort of trailer and hauling unit do you use? What are your contingencies in case of breakdowns?
- Are you willing to partition my llamas or alpacas away from other livestock? If so, will they have physical contact, while en route or unloaded at rest stops, with other lamas or with cattle, sheep, or goats? (These species have diseases and parasites in common with llamas and alpacas.)
- Will you give my lamas feed I send along with them, or do they have to eat what you provide?
- How do you deal with sick or injured animals en route?
- If something goes wrong, will you contact me so I know what's happening? Will you phone if you're running late? Will you give me your cell phone number so I can call you en route?

Professional transporters are often the best bet for long-distance moves.

And don't necessarily choose the cheapest transporter or the one who does the most advertising. Find one you're comfortable with based on your experience or other lama owners' recommendations. Transporting lamas is risky business, so find and hire the best livestock hauler you possibly can.

Choosing Your Own Conveyance

Seasoned lama travelers haul nicely in vans and SUVs. However, don't go this route unless you're sure your lamas are up to the challenge! When in doubt, opt for a traditional horse or livestock trailer and keep the following points in mind.

If you can't remove center dividers from a horse trailer you're considering, don't buy it. Lamas prefer to kush as soon as the conveyance they're riding in starts to move, and they need open space to lie down.

Good ventilation is an absolutely essential element of lama conveyance design. If you can afford it, buy an air-conditioned trailer; you will never regret it. Animals die quickly from carbon monoxide poisoning, so make absolutely certain no engine exhaust enters any area occupied by livestock.

Damp wooden flooring and some rubber trailer mats, when wet, are slick as glass. Water containers sometimes tip over, so always invest in non-slip trailer mats.

Managing Stress

Llamas and alpacas, especially when they're not accustomed to being handled

Well-trained lamas accustomed to travel can often be hauled in a van or SUV.

A Traveling First Aid Kit

Put together a traveling first aid kit to augment the kit you keep at home in the barn. Keep it in your truck or trailer at all times. If you use something from the kit, replace it as soon as you get home. Having a well-equipped first aid kit and knowing how to use it can make the difference between life and death when you're on the road and far from the closest vet. At the bare minimum, a kit should include the following:

- Several rolls of Vetwrap or a comparable self-adhesive bandage
- A roll of 2½-inch-wide sterile gauze bandage
- 1- and 2-inch-wide rolls of adhesive tape
- Telfa nonstick absorbent pads to cover wounds
- Individually wrapped sanitary napkins to use as pressure pads to stop heavy bleeding
- Antibiotic ointment
- Topical eye ointment
- Saline solution
- Betadine to flush fresh wounds
- Sterile gauze sponges
- A rectal thermometer and lubricant
- Tweezers or a hemostat
- Probios (the type labeled for ruminants) or a comparable probiotic product
- Electrolyte paste or gel
- Banamine (a prescription pain reliever and anti-inflammatory drug to get from your vet)
- Bandages and nonprescription pain relievers for yourself

or hauled, are very susceptible to stress. Stress contributes to heat exhaustion, ulcers, and a host of other serious problems. You want to minimize stress however you can, whether you haul your lamas only once—home from the person you buy them from—or you haul them many times per year.

Transport-related stress factors are of two types: short-acting factors that lead to emotional stress and long-acting factors that have physical effects and tend to accumulate over the duration of a trip.

Short-acting factors include unfamiliar surroundings, unfamiliar traveling companions, and unstable footing.

Long-acting, cumulative factors include noise, vibration, being thrown against the vehicle or other animals, insufficient food and water, and extremes of temperature and humidity.

Avoid hauling sick or injured lamas of all ages and sexes, as well as late-gestation females. Set out with sound, healthy animals, and do your best to keep them that way. A solitary lama is a stressed lama, so haul along a companion if you possibly can.

Map the route in advance. Stop-and-start driving causes hormones and blood components to fluctuate and can drive heart rates up to twice their normal rate. If the most direct route means dealing with rush hour traffic or hitting a red light at scores of stoplights, choosing a longer but easier route is a better choice.

Factor rest stops into long journeys. Plan stops along the route where you can safely offload your cargo at least once every twenty-four hours. If your lamas are seasoned travelers and you know they'll reload without a fuss, offload them every two to three hours to relieve themselves (pack along a bag of lama dung to encourage them to start a poop pile) and briefly relax.

Always carry a well-appointed first aid kit. To cut back on digestion-related illness, include enough probiotic and electrolyte paste or gel to dose each animal at least twice a day.

Load llamas and alpacas with care. The floor of their conveyance should be nonslip. If it isn't, you can cover it with several inches of damp sand. Cover all types of floors with an adequate amount of dust- and mold-free bedding. Using lots of bedding also cuts down on the vibration. To reduce noise levels, pad gates, loading chutes, and partitions with pieces of rubber matting or old blankets.

Nonslip mats should be used in hauling conveyances like this safe, open livestock trailer.

It's vitally important to keep lamas from overheating during transport. Air-conditioning, fans, and ventilation grills all make trailering safer for the lamas on board.

Allow plenty of room for each animal. As soon as their conveyance moves, most lamas automatically kush, so each llama or alpaca should be allotted enough space to comfortably lie down. Outfit larger conveyances with interior dividers that make it possible to partition individuals into compatible groupings based on sex, size, age, and disposition.

Weather extremes head the list of long-term stressors. We'll talk more about heat stress later in this book but for now, heed this warning: *it is imperative that you cool any lama suffering from heat stress as quickly as you can.* One good way to do this is to hose the animal down, saturating it all the way to the skin, using lots of cold water, and then place it in front of a fan to finish cooling off. (Do not let the lama stand in the sun, as the heat will cause the wet fiber to act as a steamer, making the animal even hotter.) When traveling in hot, steamy weather, *always* carry along coolers loaded with jugs of ice-cold water and bagged ice cubes. If a llama or an alpaca overheats, you can pack ice under it if it is kushed or hold ice against its armpits, underbelly, or groin if it is standing up.

If you don't have an air-conditioned trailer and lots of summer travel beckons, buy a power inverter that connects to your truck's battery and allows a full-size household box fan to operate on truck power. You'll need it if one of your llamas or alpacas collapses en route to a trailhead or show.

To compensate for hot, humid weather, reduce loading capacity by 15 to 20 percent (overcrowding rapidly leads to excessive heat buildup). Create additional ventilation by opening windows or replacing solid upper walls with sturdy, closely spaced pipe or heavy wire mesh. Travel only at night or during cooler morning hours, and keep the number and length of stops to the barest minimum. Never under any circumstances park a loaded trailer in direct summer sunlight.

Cold kills, too. Lamas (especially crias) are susceptible to frostbite and loss of body heat. It's important to keep in-transit animals bone dry. Cover openings to protect them from rain and wind chill, add more bedding, and allow extra space so animals can move away from chilling wind. To keep their bodies warm, crias and compromised individuals such as the elderly or fairly recently shorn should be fitted with comfy lama blankets.

Advice from the Farm

What Every Traveling Lama Needs

The experts offer tips about transporting lamas.

Time to Adjust

"Llamas' eyes take much longer to adjust to changing light conditions than human eyes do. Have you ever turned the lights on in your barn at night and watched as the llamas blink, blink, blink, blink, blink for ten or fifteen seconds until their eyes adjust? The same is true when the light level goes down suddenly. Asking them to go from a bright sunny day outside a white barn into the dark doorway of an unfamiliar trailer is asking a lot. We can see inside; they see a black hole that may not even have a floor. You can imagine the fright of being dragged into a place you can't see. They are usually not being stubborn. Look at the world through their eyes, and it will give you more patience."

—*Nancy Frank*

Comfortable Conditions

"Never tie llamas or alpacas in a trailer—it's a good way to break their necks—and always separate intact males from females. Many a cria has been created on those short hops!

"Llamas and alpacas can be easily trained to enter a van. Just make sure any holes resulting from seat removal are blocked and that you have an old rug or something down. Don't use a tarp; it's too slippery.

"Llamas and alpacas are naturally inclined to kush once the van or trailer starts moving. Sometimes packing them in for shorter time frames is actually safer for them, as their bodies will buffer each other from injury.

"On a long haul, stop every hour or two, and make everyone rises to their feet to ensure that they don't lose circulation due to being kushed for too long."

—*Deb Logan*

A Buddy and Some Elbow Room

"Alpacas can be transported in a minivan or small trailer because they will kush during transport. They do *not* like to travel alone, so if you have to take one in an emergency situation, take a buddy as well to help relieve stress. But just because alpacas can be transported in a small space doesn't mean they should be. Try to transport in a safe environment with as little stress as possible."

—*Tina Cochran*

Allow sufficient time to drive carefully. Accelerate and brake slowly and smoothly. Ease up on the gas well ahead of turns, and don't take corners too abruptly. Factor in load checks, too; stop twenty minutes after departure to check your load and at least every hour or two after that. Allow your lamas to stand up during these checks to keep circulation healthy.

If you make your lamas' first rides physically comfortable and relatively stress free, they'll learn to take traveling in stride—and so will you.

Finding and Working with a Veterinarian

Line up a veterinarian or two before bringing your animals home. The best way to find a *good* one is to ask other lama owners in your locale for recommendations. Narrow it down to two to four veterinarians most people like, then call their offices and ask the following questions.

How many veterinarians are associated with your practice? Are all of your veterinarians familiar with the care and treatment of llamas and alpacas? Can I stipulate which one I want to see my animals?

Do your veterinarians make routine farm visits, or do I need to bring my animals to your clinic for everyday procedures? What facilities are available if I have to leave them in your care? Will someone come to my farm in an emergency? What about after-hours, weekend, or holiday emergencies? To whom do you refer clients when you're unavailable?

If I phone your clinic with a problem, will you connect me with a veterinarian (if one is there), or will you relay my concerns and return my call?

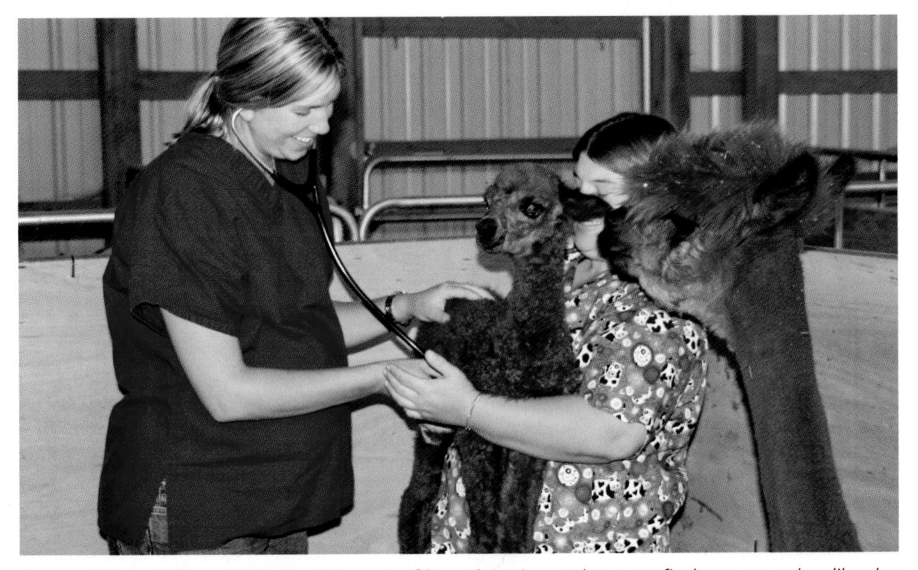

A good lama vet is an indispensable part of lama keeping. When you find one, treat her like the treasure she is!

How is payment handled? Is it cash up front for every call, or do you carry a tab? Are credit cards acceptable? Do you offer payment plans?

Visit the Clinic

If you like what you hear, arrange for a time to visit the clinic. When you do, does the staff seem friendly and knowledgeable? Will they let you speak with a vet? Desk personnel and vet techs are your link to working veterinarians, so if they make you feel uncomfortable, keep on shopping.

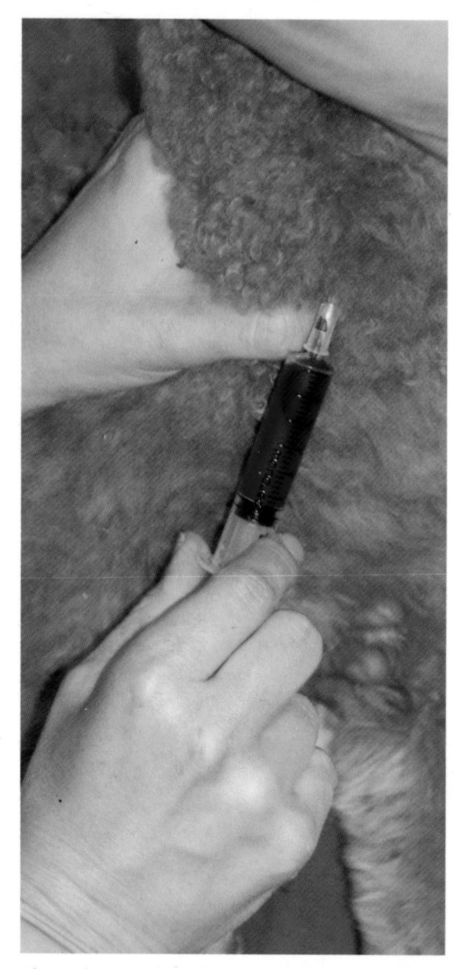

If you have a lot of lamas, you may want to ask your veterinarian to teach you to do routine vet work such as giving shots and drawing blood.

Check the boarding facilities. Are they safe, clean, and arranged so patients can't physically interact with one another? What sort of feed are the animals eating? Can you provide your own feed if you want to? Is water readily available, and is it clean?

If bedside manner matters to you, do you like the veterinarians you'll be working with? What is the practice's take on clients performing routine health care procedures, such as treating minor injuries and giving their own shots? Once the vets you work with realize you're competent, will they dispense prescription drugs such as Banamine and epinephrine?

If the vets aren't particularly lama-savvy, will they mind if you research a problem in books or online and bring them your results? Some appreciate input; others take offense.

Whatever you ask of the veterinarian, you should never be made to feel inadequate or stupid. They're your lamas and you'll be footing the bill, so if you're made to feel small, take your business elsewhere.

Schedule a Routine Farm Visit

Once you've selected a practice, don't wait for an emergency to use its services. Schedule a routine farm visit, and see how that goes. Does the veterinarian arrive promptly, or does someone from the office phone to inform you of delays? How does the veterinarian interact with your llamas and alpacas? Are you comfortable with the vet's attitude and work?

In turn, it's only fair to treat your new veterinarian right. Have your llamas or alpacas caught, cleaned up, and ready for treatment when the vet arrives. Chasing your animals across a 40-acre field is not the vet's job.

General Advice

Learn to handle emergencies until your veterinarian arrives. Make certain your cell phone or telephone works in the barn in case your veterinarian needs to talk you through a procedure. Stock well-equipped first aid and birthing kits, and know how to use them.

Never wait until a minor problem escalates into an after-hours or weekend emergency. Know what you can and cannot do by yourself, and involve a veterinarian as soon as one is needed.

Be there when your vet works with your lamas. Your animals know you, and they'll behave better if you're on hand to help. If you don't understand a treatment, ask questions. If follow-up care entails detailed instructions, write them down and follow them to the letter.

Furnish a comfortable, weatherproof, well-lit place for your veterinarian to work. Provide a reliable means of restraint even if it's a makeshift chute or extra hands to help steady the patient securely against a wall.

A cold beverage on a sweltering summer afternoon or a steaming cup of coffee in the winter is always appreciated. And always settle your bill when payment is due.

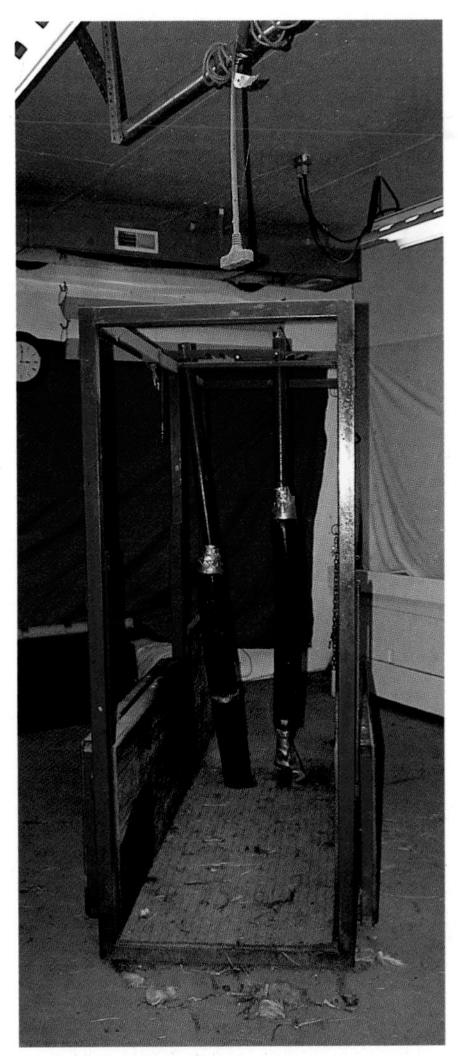

A strong, safe handling chute like this one makes veterinary procedures easier for everyone concerned.

CHAPTER THREE

Handling Llamas and Alpacas

Many potential lama owners see ads depicting cherubic children hugging cuddly alpaca crias, and they think, "That's for me!" What could be more appealing than a winsome llama or a snuggly alpaca to cuddle and smooch? Unfortunately, it isn't usually that way. In fact, cheeky crias that instigate lots of human interaction usually grow up to be spitty, pushy females or aggressive, out-of-control geldings and males.

Llamas and alpacas can be taught to enjoy human contact, but it's not something the average lama seeks. If you observe groups of lamas for a while, you'll notice they don't scratch each other's itchy places in the manner of horses, nor do they cuddle with one another to sleep the way goats do. Once past the stage when crias stay close to their mamas, lamas generally touch each other only when males play fight. Touch isn't something they expect from one another, so it's alien to their sensibilities when humans want to touch them.

Like wary cats, lamas like what they like, when they like it. You can't push yourself on llamas and alpacas and expect them to reciprocate. If you keep that in mind and are patient, most lamas will eat from a bucket or the underside of a Frisbee you are holding and will learn to be haltered and led with a minimum of fuss. Lamas are intelligent and forgiving, so once they trust you, even if they've suffered injustices in the past, they happily submit to further training—as long as you treat them fairly. If you want an in-your-face, "Pet me! Pet me!" animal friend, buy a dog or a goat.

Llamas and alpacas don't respond to human intervention in the manner of any other livestock species, so understanding lamas before you begin handling them is crucial—for us and for them.

Understanding Lamaspeak

To further understand your llamas and alpacas, you need to learn lamaspeak. To do that, get yourself a soda or a glass of lemonade, and take a comfy lawn chair out where your lamas are; sit down and observe as they interact with one another. Spend an hour or two a week watching lamas, and soon you'll be an expert in their language. Below are some of the sounds and actions you're likely to hear and see.

Lama Vocalizations

Although the Quechua and the Aymara refer to llamas and alpacas as "our silent brothers," lamas really aren't that silent. They have a wide array of vocalizations, which you will come to recognize as you get to know them. Here are a few to listen for.

Humming

Humming is the primary way lamas express themselves. New owners often think lamas hum because they're happy or content, but that is generally not the case. Lamas usually hum at various volumes and pitches when they're curious or feeling cautious, bored, lonely, worried, overheated, cold, in pain, frightened, or distressed. However, mothers do croon a quieter, more peaceful hum to their crias, and the crias hum softly back.

Clucking and Snorting

Lamas cluck to intimidate another lama, when telling another lama to "leave me alone," or sometimes when they're simply worried or concerned. Females cluck when their babies wander too far away or if another mom or cria ventures

Llamas and alpacas are herd animals that need other animals' company to avoid feeling stressed.

too near. Lamas snort in response to the same stimuli, and clucks are sometimes interspaced with snorts.

Grumbling

Lamas grumble, a sound made deep in their throats, when issuing a fairly mild warning to other animals or when annoyed.

Alarm Calling

When one lama perceives danger, it issues a distinctive alarm call to alert the rest of the herd. The alarm call is a loud, undulating whinny that some owners compare to a turkey call. When you hear it, you'll recognize it, believe me. When lamas repeatedly scream an alarm call, investigate. They might not be in danger, but they certainly think they are. Some lamas sound the alarm more often than others; one of the females who lived on our farm called loudly, stridently, and continually every time our pet pigs roamed the yard!

Screaming

Lamas scream when they're truly angry—it's a sound you simply can't miss. Males scream when fighting one another or when warning another male away from their territory. Lamas, especially alpacas, also scream when they're deeply frightened or stressed.

Orgling

A male lama orgles when mating with a female. It's a unique, throaty sound that varies greatly from individual to individual, but it always indicates mating activity. When one male orgles, all of the other intact males in the area sometimes join in. Males orgle throughout the breeding process, from courtship through the final act, sometimes for as long as an hour at a time.

Did You Know?

A fun way many llama and alpaca owners interact with their lamas is by encouraging "lama kisses." Although lamas don't particularly like to be touched, they are curious animals and will reach out their necks from a respectful distance and, if you let them, gently sniff or touch your face. To encourage this, lock your hands behind your back and lean forward to meet the lama partway.

Body Language

Lamas have a wide repertoire of body movements and stances that alert a savvy owner to what's going on. For instance, when a lama stands rigidly erect, ears pricked forward, with tail raised, it is focusing on a distant scene or object of concern. Sometimes entire groups of lamas stand like this until one

This gorgeous Klein Himmel Llamas male is on alert. "What's going on?" he wants to know.

This llama's posture clearly says, "Back off or you are in deep trouble!" Next, he will spit.

member decides the scene or object is scary and calls an alarm; the others then commence calling or fleeing. Or the sentinel member decides what it is seeing is benign and interesting and relaxes or approaches the object in question, signaling to the others that the object is benign; the rest of the lamas then forge forward to investigate en masse.

Intimidating Poses

When a male or an assertive female strikes a broadside pose with neck arched, ears pinned back, and nose and tail held high, it's saying, "This is my territory; back off!" Because this is an intimidation pose, it's important that you do not walk around the pushy lama. Instead, make it move so you can walk straight ahead. This is a simple way to tell that lama that it's not the boss of you!

When two lamas face each other, bodies rigid, ears pinned back, heads up and chins elevated, and tails held high, they're engaged in a lama standoff. This usually occurs between fairly equal-ranking members of a herd. In most instances, one will eventually turn its head or walk away. If not, further animosities will ensue, such as spitting, jostling, and verbal threats.

Low-ranking herd members, particularly crias, indicate submission by lowering their heads, necks curved, and flipping their tails across their backs.

Spit Happens

Lamas spit at one another for various reasons, including to establish dominance over a herd mate, to indicate extreme displeasure or intense fear, and even to discipline their crias. The average lama, however, rarely spits at human beings.

Anytime a lama tenses, pins back its ears, and elevates its chin, it is probably annoyed, and spit could be in the offing. Spit, however, happens in degrees. The mildest display is an air spit; the lama simply makes a spitting sound but nothing comes out except an explosion of air. Air spits means, "Leave me alone!" or "I'm getting scared (or mad)."

The next level is a full mouth spit. This sometimes happens if a lama air spits while it has something in its mouth; the full mouth spit is usually more forceful than a simple air spit. The lama spits saliva, chewed grass, or dry feed. This means, "I told you to back off! If you keep this up, you'll be very sorry!"

Finally comes the true, regurgitated from the depths of the stomach, spit that no animal or human wants to have spat upon him. Regurgitated stomach contents are slimy and smelly beyond words; even the spitting lama knows this and saves it for a last resort. After an incident involving regurgitated stomach contents, both parties, the lama that spits and the one spat upon, generally spend fifteen minutes to half an hour with their mouths hanging open in response to the terrible taste and smell. Deep stomach spit is washable, and it certainly won't hurt you, but if you can avoid it, you'd be very wise to do so.

Borrowing Lama Body Language

Once you understand how lamas think, you can borrow their body language to suit your needs. For instance, many lama owners get results when they air spit back at spitty lamas (just be certain your spit is the final one in the exchange). Some owners establish dominance over

These llamas are threatening one another with spit. Notice that the black one is standing broadside to show her adversary how important she is.

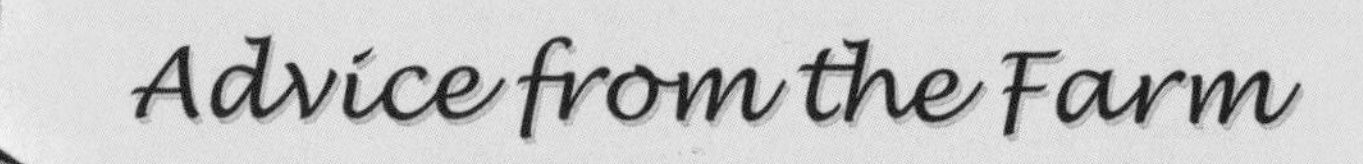

How to Handle the Slimy Stuff

The experts tell us what to do when "spit happens!"

Make Him Smell It

"Alexander is so easy to handle most of the time, but when we shear him, oh boy that's something else. He hates it and he spits—a lot! A woman on one of the email lists I subscribe to suggested tying a bandana loosely around Alex's nose. That is, fold the bandana into a triangle and then place it over his face below his eyes with the point of the triangle hanging over his nose and the sides tied on to his halter. Then, he can breathe easily, but when he spits—and it's that green slimy stuff right from his stomach—some of it gets caught in the bandana, and he has to smell it. He hates that, and he usually only spits once or twice. And it doesn't land on us!"

—Jan Johnson

Duck!

"Spitting is an alpaca's only defense. They use it on each other; you just need to stay out of the line of fire. Tip: Don't get between two alpacas at feeding time, and if you have to, when you hear it coming, DUCK!"

—Tina Cochran

Shout It Out

"Oftentimes, when you see them bring up a cud and start chewing in preparation for a spit, a simple hand in front of their face combined with 'No spit!' will do the trick."

"Or, blindfold the animal. If he's historically 'spitty,' use a piece of sheet or other lightweight fabric, and let the ends hang down over his nose and mouth.

"You could also buy a spit mask. They're made of mesh and therefore lightweight and easily added to the llama's ensemble—just slip it over the nose and buckle it behind the head. Hose it off between uses, and if it gets too awful, wash it with your socks!"

—Deb Logan

Get It Off

"If you do get seriously splatted with spit and can't get the stench off of your body or out of your clothing, try scrubbing them with a mechanic's hand cleaner like Mean Green, Goop, or GOJO. These zap even buck goat scent, so removing lama spit stench is easy!"

—Sue Weaver

Pee-yoo! After a spitting match, both the spitter and the spitee air their mouths for up to half an hour.

a pushy lama by returning the rigid, chin-up stance that is part of a lama standoff until the animal backs down. Just remember that if you choose to communicate with llamas or alpacas in lama body talk, you must not allow a pushy lama to win.

Other Significant Behaviors

Llamas and alpacas are herd animals; they never like to be alone. This is why it's kindest to buy or adopt two lamas instead of just one. The exception would be guardian llamas kept to watch over sheep or goats. These llamas bond with the species they guard and accept the other animals as their own.

A solitary lama is likely to suffer from stress. Other stressful situations for lamas include being caught when they aren't used to it, receiving a toenail trimming, being shorn, being separated from a favorite friend or subjected to unpleasant veterinary procedures, and being hauled. Signs of stress include humming, breathing open-mouthed, pacing, laying ears back, tensing lips, the appearance of "worry wrinkles" below the eyes, swishing the tail, and refusing food or water. Stress leads to any number of unpleasant medical conditions, so avoid stressing lamas whenever possible.

Lamas urinate and defecate on communal dung piles. They create a number of potty piles in their housing and grazing areas and return to them whenever nature calls. The lamas don't soil the areas where they sleep and graze. All of this makes for easy cleanup. Some golf courses allow golfers to use llamas as caddies, and pack llamas are often welcome on state and national trails where messier, more destructive pack animals are verboten.

Because of this behavior, lamas can be hauled in vans and SUVs and even house-trained should you choose to keep unusual household pets. The downside is that they may not eliminate away from home if they can't find a proper potty pile, so when you travel,

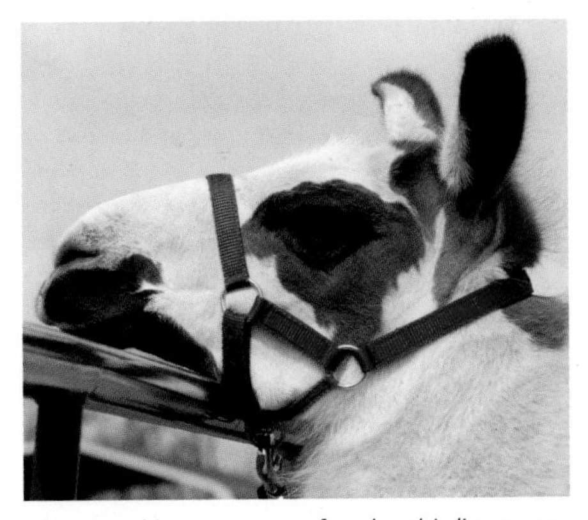

A worried lama's eyes soften but his lips tense. The wrinkle under Bandit's eyes indicate he's concerned about being groomed for the very first time.

Even the youngest lamas quickly learn to use the dung pile.

pack along a can of "lama beans" from home to start a new pile when your lamas need one.

Llamas and alpacas also create "dust bowls," where they roll to fluff up their fiber and maintain its insulating qualities. Sometimes after or during rolling, they lie in their dust bowls, flat out on their sides or on their backs with their tummies to the sun. Don't worry, this is normal lama behavior.

Aberrant Behavior Syndrome

Something all newbie lama owners should be aware of is aberrant behavior syndrome (ABS). The spitty, aggressive llamas and alpacas we mentioned in the opening paragraphs of this chapter are probably ABS lamas. It's a serious, growing concern in the lama world. Here's what you need to know.

What Causes ABS?

The following cautionary tale originally appeared in the March/April 2008 issue of *Hobby Farms* magazine.

While strolling through the flea market or animal swap, you spy that most winsome and wonderful of baby creatures—an adorable infant cria. Its huge eyelashes, soulful eyes, and fuzzy coat steal your heart. The seller assures you that raised on a bottle (and he'll supply the bottle), Kuzco will be as tame as a kitten and as loyal as the family pooch. You peel off a fistful of twenties, scoop up Kuzco, and head for home with your gorgeous new pet.

And the seller was right! The family hugs and kisses and coos at Kuzco, the kids tussle playfully with him, and he grows up adorably sweet and well loved.

Then something happens. Just a few months before Kuzco's second birthday, you enter his enclosure expecting the usual warm welcome, and he charges you and rams you with his chest and knees. You're shaken, but this is Kuzco—he'll get over it, right? But he doesn't. Kuzco begins screaming and spitting at people on sight, and he charges and stomps anyone foolish enough to enter his pen. One day he knocks you down and really works you over. He's dangerous, that's for certain, but what to do? You begin calling llama and alpaca breeders, and they say the same thing: Kuzco is a victim of ABS, and there is no sure cure. Unless you can find a rescue willing to accept and retrain Kuzco, you'll have to have your boy put down.

Overhandling causes baby lamas to imprint on humans, which is what triggers ABS. Males are more likely to develop it, but females can be dangerous, too. ABS lamas share some or all of the following background:

- They were separated from their mothers at an early age and bottle raised.
- They were raised apart from other lamas (particularly adult role models).
- They were handled by people unaware of normal camelid behavior. (They were raised by inexperienced owners, displayed in petting zoos, or cared for by vet clinic personnel unaccustomed to interacting with lamas being boarded for medical concerns.)

So unless you're camelid-savvy, don't try to bottle-raise a cria. If you must do so for any reason, contact an experienced breeder or a bona fide lama rescue for advice. Bottle raising a lama can be done, but it's not like bottle raising a calf or a lamb.

Is It ABS or Not?

Not every animal that behaves in an erratic manner is an ABS lama. Untrained lamas are afraid of human contact and react, sometimes violently, when subjected to frightening procedures, such as being cornered and caught at toenail-trimming or shearing time. According to the experts at Southeast Llama Rescue (see the Resources section), you can expect untrained lamas do the following:

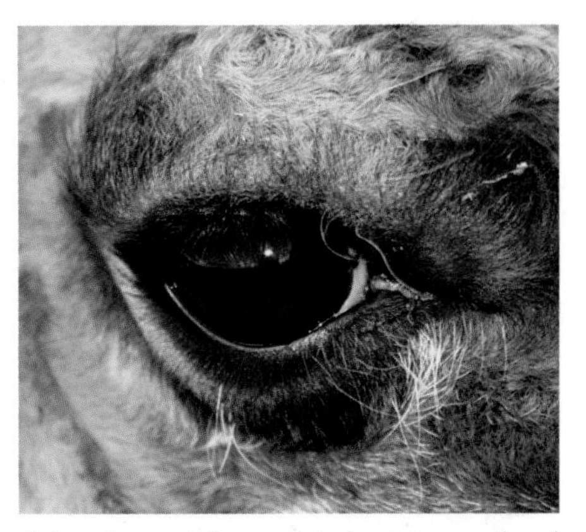

A lama's eyes tell you a lot about its mood and well-being.

- Swing their rumps or shoulders into people while trying to avoid being touched.
- Run into people with their chests when trying to evade capture.
- Swing their necks and heads around, especially when being haltered or touched on the head or neck.
- Kick.
- Spit.
- Fall on a person who is trying to pick up a foot.
- Spook at new stimuli while being led, and drag anyone holding on to the lead rope.

Did You Know?

Some intact male lamas misinterpret your intentions if you bend over in front of them, perhaps to pick up something from the ground, or if you squat to fix the lower rail on the fence. They think you've kushed because you want them to help you make a cria, so watch your back when working around the guys!

A female llama issues a vocal warning.

There is nothing wrong with these lamas that patience and training won't cure; they react the way they do instinctively, out of fear.

ABS lamas, however, are highly aggressive. Although they are dangerous, with careful retraining by handlers who are wise in the ways of these aggressive lamas, many can be saved. Early signs of aggression in a lama include these behaviors:

- Following people within a two-foot distance without enticement such as food.
- Greeting people by flipping its tail across its back and curving its neck into a U-configuration.
- Positioning itself to prevent people from passing by.
- Orgling (males) at humans.

If aggression isn't addressed early on, behaviors intensify and may escalate to the following:

- Screaming at or spitting on people without provocation.
- Charging at and jumping over or trying to crawl through fences to reach bystanders.
- Charging into humans, ramming them with knees and chest, and possibly stomping them after they've been knocked down.
- Rearing.
- Biting.

Can an Aggressive Lama Be Saved?

Suppose you've unknowingly purchased an aggressive llama or alpaca—can it be saved? In many cases, yes, but you'll need the advice of experienced rehabilitators to safely handle the task. Before you try, contact Southeast Llama Rescue for advice; they've successfully helped to rehabilitate many aggressive llamas and can help you formulate a plan.

Since most aggressive/ABS lamas are intact, hormone-crazed males, the first step is have your lama gelded, then give him a few months for his hormone levels to subside. According to Southeast Llama Rescue, gelding, followed by training, is successful a majority of the time.

Prevention, however, is always better than cure. When raising crias, don't encourage overfamiliarity. It's all right to interact with youngsters, but don't allow cheeky individuals to mouth your clothing, fiddle with your shoes or shoelaces, sniff your crotch, rub up against you, push into you, or wrap their necks around you. Adult lamas don't allow these brazen liberties (all of which are precursors to later aggression), and neither should you.

The Kindest Cut—Gelding Male Lamas

The best thing you can do for male lamas you don't plan to use for breeding is to have them gelded. Since very few llamas and alpacas are herd sire quality, this means most of the males you're likely to know.

Geldings, especially when castrated at the proper age (most veterinarians say twelve to eighteen months), are far less likely to develop aggressive tendencies than their intact brothers are. They can be pastured with other llamas and alpacas, increasing their social opportunities and thus their happiness manifold, and they are easier and safer to handle and train. Another benefit: alpaca and long-wooled llama geldings often produce finer fiber than intact males do.

In most cases, gelding before twelve months of age is not recommended because early castration can lead to long bone growth abnormalities. Most veterinarians agree, however, that early castration of dangerously aggressive young males is by far the lesser of two evils.

Lamas are usually gelded using one of two techniques: scrotal castration (in the manner of castrating stallions and boars) and prescrotal castration (the method used to castrate dogs).

Scrotal castration can be done with the animal standing or lying down and using local or general anesthetic; prescrotal castration is always performed with the patient lying

Lamas, such as these attractive alpacas, appreciate low-key handlers who communicate with them in body language they understand.

Normal intact testicles. Only the best stud male prospects from every cria crop should remain intact. Geldings make great hobby farm fleece producers and pets!

down. Some veterinarians close the area with sutures, some don't.

Food (but not water) should be withheld for twelve hours prior to surgery, and newly castrated llamas and alpacas should be observed closely for twenty-four to forty-eight hours post-surgery. The incision site needs to be monitored until healed. Problems to watch for include unusual swelling, bleeding, and discharge of any sort, fly infestation, and difficulty urinating. Postoperative problems are rare, but if in doubt always call a veterinarian.

Working with Lamas

Dress properly before setting out to work with llamas or alpacas, especially if they're largely untrained. Wear sturdy shoes in case a lama treads on your foot with its hard toenails (that hurts!), gloves to protect your hands from rope burns, and safety glasses or other protective eyewear if you're dealing with a lama that's likely to spit.

Herding Lamas

Unless you have remarkably well-trained lamas, you'll probably need a catch pen or even a chute to halter your animals. (We'll talk about catch pens and chutes again in chapter five.) You'll also need a pair of 4-to-5-foot herding poles to act as arm extensions. (Make your own, order CAMELIDynamic wands, or buy white cattle fitting poles at well-stocked farm shops.)

To herd lamas, apply pressure by quietly walking toward them from the opposite direction with your herding poles extended out to your sides. Stay to the rear and slightly to one side when moving a group of lamas (they tend to stay together as a group when being driven), and don't shout or make a fuss. Just quietly herd them in the direction you want them to go. If any seem inclined to leave the group, move from side to side to keep them moving together.

When you've contained them in a smaller area, direct the individual you want into a catch pen or chute using the same techniques.

When working with lamas, remain calm and project a relaxed, peaceful attitude. Avoid fast, jerky movements, and don't make direct eye contact; this upsets them.

The Importance of Proper Halters

If you want happy, well-behaved llamas or alpacas, get them the right kind of halter. Consider this: lamas are semi-obligate nasal breathers. This means they can't fully breathe through their mouths, so anything that impedes airflow through a lama's nostrils—like the noseband of an ill-fitting halter—is very, very scary to the lama and, in extreme cases, life-threatening. When a tight or improperly adjusted halter noseband slips down off a lama's short nose bone, he's going to be frightened and respond by acting out.

Halters come in three styles:

- Fixed noseband halters, which have adjustable crownpieces (the straps that pass behind the lama's ears) but nonadjustable nosebands.
- X-style halters, which have a crownpiece and a noseband that form a continuous loop.
- Fully adjustable halters, which have lots of noseband and crownpiece adjustments.

Fully adjustable halters are by far the top choice because they can be custom fitted to your lama's head, so the animal is less likely to become anxious and misbehave. Two of the best adjustable lama halters on the market are Marty McGee Bennett's Zephyr halter and Cathy Spalding's Gentle Adjustment halter. (See CAMELIDynamics and Gentle Spirit, respectively, in the Resources section.) However, any fully adjustable halter with short cheekpieces (the part of the halter that connects the noseband to the rest of the halter) works. Short cheekpieces help keep the noseband high on the lama's face—which is where it belongs.

This pretty suri llama wears a properly fitted, nonadjustable noseband halter.

Stormy is wearing Bandit's large-size Zephyr halter, adjusted properly for his smaller head.

John guides Stormy into the squeeze chute using two herding poles.

When you've isolated the animal you want, work it into a corner or chute, then calmly place your hand on the lama's back to steady it. When working in a catch pen where the animals are free to move about, remember that lamas instinctively oppose direct pressure. If you push against one, it'll probably push back. Give the lama time to relax, then calmly slip the halter on to its head and adjust the fit.

Llamas and alpacas are quick learners and easily conditioned; if you're patient and persistent, catching them gets easier each time.

The Lama Whisperers

If you're really interested in learning to work with llamas and alpacas in a gentle, effective manner that they truly understand, check out the Web sites of lama clinicians Marty McGee Bennett (CAMELIDynamics), Cathy Spalding (Gentle Spirit), and John Mallon (the Mallon Method), where you can pick up tips and buy educational DVDs and the equipment you need to train your animals using these proven techniques. All of these clinicians are listed in the Resources section.

Advice from the Farm

To Catch a Lama

The experts talk about haltering and catching.

Ask, Don't Tell

"Llamas and alpacas can't focus on things that are much closer than 10 or 12 inches, so a halter coming at them right by their noses is startling. Starting it out as far away from their noses as you can reach and *asking* them to put it on will be much easier in the long run."

—Nancy Frank

Go Verrry Slooooowly

"To catch a llama (or alpaca), first try to herd it into a catch pen no larger than 12 foot square. We use a 10 x 10 foot permanent enclosure for shearing and vetting as well as a portable 8 x 8 foot enclosure constructed of aluminum panels for basic training or haltering. A catch pen is a nonnegotiable requirement for adoption through Southeast Llama Rescue, so that should tell you how important we feel it is. Most animals cannot be haltered in the field, especially if there is any panic in the air. A catch pen is a great vehicle for facilitating all types of training and, of course, drama-free haltering.

"Llamas and alpacas typically stop in their tracks if you stand by the animal's shoulder facing the same direction he is. Slowly fasten a lead or rope high on his neck, right below his head, to secure him for haltering. Go slowly—never grab. Speak softly and calmly. Did I say 'go slowly'? Go verrry slooooowly, and breathe!

"Make sure the halter you use is the appropriate size and that the noseband is sturdy enough that it can be held open without collapsing. If push comes to shove, too big is better than too small. If a llama feels as though he's suffocating, the rodeo will have just begun!

"Remember that llamas and alpacas are naturally head shy, so don't reach over their heads. Rather, bring the halter up from underneath."

—Deb Logan

CHAPTER FOUR

Feeding Llamas and Alpacas

Feeding llamas and alpacas is a complex subject. It's best to discuss the matter with your county extension agent or a livestock nutritionist. This person can suggest forage, concentrates, and supplements suited to your climate and your herd's needs, which are based on the animals' ages and uses, while taking into consideration the types of feed available in your locale. However, certain truths apply no matter where you live or what sorts of llamas or alpacas you feed. We'll discuss those in this chapter.

Whatever you're feeding your lamas, dietary changes must be made over a period of time; ten days to two weeks is sufficient. Abrupt changes trigger potentially life-threatening digestive upsets. This is important; I simply can't stress it enough. It's also important to establish a routine and stick to it. Don't stress your llamas and alpacas by skipping or delaying their meals. Make certain each animal is eating; lamas that refuse their usual feed are probably ill.

Ruminate on This

Lamas are modified ruminants. Their forestomachs are made up of three compartments rather than the true ruminants' (sheep, goats, cattle, deer) four. Whether three- or four-compartmented, ruminants' digestive systems are very unlike those of simple-stomached species such as horses, carnivores, and humans. The three sections of llamas' and alpacas' forestomachs are called C-1, C-2, and C-3; each compartment has a specialized job to perform.

C-1, located on the animal's left side, is the largest (and first) compartment; it makes up roughly 80 percent of the stomach's total volume. C-1 secretes no

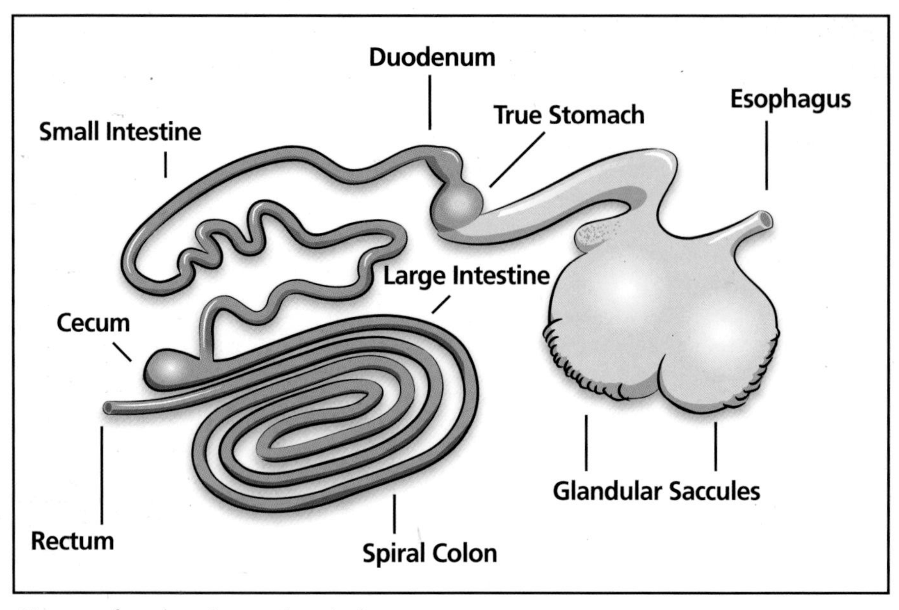

Diagram of an alpaca's gastrointestinal tract.

digestive enzymes—it's essentially a fermentation vat housing a horde of friendly microorganisms that convert cellulose into digestible nutrients. Newly eaten feed liberally mixed with saliva comes into C-1 by way of the lama's esophagus, then fermentation begins. Coarse bits of feed are periodically regurgitated, rechewed, and then reswallowed, a process known as rumination or chewing the cud. Additional chewing reduces particle size and churns in more saliva, important because saliva adds bicarbonate and phosphate buffers to combat acid production during fermentation. The average healthy llama or alpaca ruminates about eight hours in a twenty-four-hour period.

Ingested material stays in C-1 for roughly sixty hours, where it's continually mixed by strong, spontaneous contractions of the forestomach. The material next moves into C-2, where some absorption of nutrients occurs, and then on it goes into compartment C-3.

C-3 is a tubular organ running alongside C-1 on the right side of the abdomen; it holds 11 percent of forestomach volume. The last one-fifth of this tube contains true gastric glands, so C-3 is sometimes called the true stomach. Stressed llamas and alpacas frequently develop ulcers in C-3. (We'll talk about gastric ulcers again in the appendix.)

Further digestion occurs in the small intestine. Material then presses on to the cecum and spiral colon, where vitamins, minerals, and water are absorbed and fecal pellets are formed of the remaining waste and eventually eliminated.

Quality grass or grass hay should form the backbone of every lama's diet—it's what their digestive systems evolved to process. High-protein alfalfa and rich concentrates contribute to obesity, acidosis, and bloat, so feed them in moderation if at all.

In newborn crias, only the true stomach is fully functional. However, as it suckles its dam and begins nibbling plants, a cria ingests the microbes needed to kick-start C-1 function. By eight weeks of age, the lama's C-1 reaches adult size, and by twelve weeks, the animal is functioning like a grown-up lama.

Lamas, particularly llamas, are far more efficient rough pasture feeders than are cattle, sheep, and even goats. Lamas consume 20 to 40 percent less feed per unit of metabolic body weight than sheep do on the same diet, primarily because llamas and alpacas produce more saliva in relationship to foregut volume than sheep and goats do. The pH of C-1 is closer to neutral, which favors cellulose-friendly microbes and enhances fiber digestion. Digestive matter remains in C-2 longer, allowing microbes to process more fiber; blood nitrogen is extracted from the kidneys and used more efficiently in llamas and alpacas; and liquid passes more rapidly through the camelid gut.

The ABCs of Feeding Lamas

The basic elements of every lama's diet are pasture, hay, concentrates, minerals, and water. Following is some essential information on these nutrients.

Poisonous Plants

No matter where you live in the United States, there are probably poisonous plants growing where your llamas and alpacas graze. This may or may not be a problem, depending on the following factors:

Whether your animals ate them. Poisonous plants aren't necessarily attractive to llamas and alpacas. It's said that pasture-wise animals of all sorts seem to intuitively know which plants they can safely consume (but don't count on it). And many poisonous plants taste nasty—either acrid or extremely bitter—so lamas won't eat these unless they're hungry and it's "eat them or starve."

Whether they ate enough of the plant to matter. Many so-called poisonous plants are simply toxic. Unless they're eaten in massive quantities or over a length of time, they do no harm.

Whether they ingested the poisonous part of the plant. In most cases, only a portion of a plant is poisonous, such as its roots or wilted leaves or seeds. Or the plant is poisonous only at certain stages of its growth, and animals don't eat it at that time of the year.

Whether they're immune to the compounds in a given plant. Many poisons are species specific. Although not a lot of research has been done into lama-poisonous species, at least some part of the following wild and domestic plants and trees are known to be toxic or poisonous to true ruminants such as cattle, sheep, and goats. It's best to keep them out of your pastures and away from lamas.

• • •

To prevent inadvertently poisoning your llamas or alpacas, plant vegetable and flower gardens where lamas can't reach across fences and partake of their bounty, and dispose of garden clippings, garden refuse, and yard waste well out of reach. Never add garden cleanings to lama feed!

Keep your pastures and fence lines free of suspect plants, including trees known to poison similar livestock. Grub them out if you can; herbicides alter the taste of some poisonous plants and unfortunately render them lama tasty.

If you suspect a llama or an alpaca has ingested a poisonous plant, call your veterinarian without delay. Don't wait to see what happens; animals die quickly from many poisons.

Plant Species Poisonous to Llamas and Alpacas

Aconite
Amaryllis
Arrowgrass
Avocado
Azalea
Bagpod
Baneberry
Barberry
Belladonna
Bellyache-bush
Bittersweet
Bitterweed
Black locust
Black snakeroot
Black walnut
Bleeding heart
Bloodroot
Blue cohosh
Blue flag
Boxwood
Broccoli
Broom corn
Brussels sprouts
Buttercup
Cabbage
Celandine
Cherry
Chokecherry
Cocklebur
Coffeeweed
Cowbane
Cowslip
Crocus
Crow poison
Crowfoot
Crown of thorns
Daffodil
Deadly nightshade
Deathcamas
Devil's ivy
Devil's weed
Dogbane
Dogtooth lily
Doll's-eyes
Eggplant
Elderberry
Elephant ear
False hellebore
Greasewood
Hemp
Holly
Horse nettle
Horsetail
Hyacinth
Hydrangea
Indian hemp
Indian poke
Inkberry
Iris
Jimsonweed
Johnson grass
Jonquil
Kale
Klamath weed
Labrador tea
Laburnum (golden rain)
Lantana
Larkspur
Laurel
Lily-of-the-valley
Lima beans
Lobelia
Locoweed
Lupine
Marijuana
Milkweed
Monkshood
Moonseed
Mountain laurel
Nightshade
Oak
Oleander
Onions
Philodendron
Poison hemlock
Poke
Poppy
Potato
Privet
Ragwort
Rattlebox
Rattleweed
Rhododendron
Rhubarb
Rock poppy
Rusty-leaf
Senecio
Sesbania
Snakeberry
Sneezeweed
Spindle tree
St. John's wort
Stagger grass
Staggerbush
Star-of-Bethlehem
Tansy
Tansy ragwort
Thorn apple (common name for jimsonweed)
Tomato
Velvet grass
Water hemlock
White cohosh
White hellebore
Wild black cherry
Wild parsnip
Wisteria
Wolfsbane
Yellow flag
Yellow jasmine
Yew

Pasture

Llamas not only graze but also browse. That means they'll rid your pastures of a certain amount of brush and unwanted plants, such as blackberry and wild rose, while nourishing themselves. Alpacas browse, but not as efficiently as llamas do, and they prefer (by far) nice soft grass. You can probably get by with existing pastures when adding a few lamas to your hobby farm menagerie, but if your plans exceed a few head, discuss pasture renovations with your county extension agent. The agent can tell you which grass and legume mixes thrive in your locale and how to overseed them into existing pasture. If you need to start over from scratch, the agent will help you with that option as well.

Pastures seeded specifically for lamas are usually mixes of several grasses and possibly a legume or two. Orchard grass, timothy, fescue, white clover, and alfalfa are typical ingredients. Keep in mind that some types of pasture forage must be reseeded at intervals, while others just grow and grow.

Hay

High-fiber llama and alpaca diets based on high-quality pasture and hay are best. The best dry forage is long-fiber grass hay. Feeding too much high-protein hay such as alfalfa, clover, lespedeza, and other legumes can create the same problems as high-protein concentrates do. (Because hay is so important, we'll discuss it in more detail in the next section.)

To prevent unpleasantness at feeding time, provide enough feeder space for everyone to eat at the same time. This safe, homemade feeder is perfect.

When feeding hay, keep in mind that many llamas and alpacas are selective eaters. They nibble choice bits of hay and dump less savory morsels on the floor or ground, where they'll eventually be trampled. To save yourself money and aggravation, feed hay from waste-resistant feeders, using any discarded hay for bedding (or feed it to less picky species such as cattle and horses). Allow enough hay rack and feeder space for every lama in a group, even the shy ones, to comfortably eat together.

Keep feeders *clean*. Most llamas and alpacas won't (and for health reasons, shouldn't) eat or drink from fouled hayracks, feeders, and water sources. This is especially important if messier species, such as equines or cattle, share your lamas' living space. (See chapter five for more information on feeders.)

Concentrates

Concentrates such as grains and commercial feeds based on grains ferment more rapidly than forage does, so they produce excess acid that can kill microbes and eventually the animal they serve. Therefore, it's important to feed concentrates only when individuals truly need it. Nutritionists say that late-gestation females usually require grain at the rate of 1 percent of their total body weight, and lactating females and growing youngsters at 2 percent of total body weight.

Your best bets if you feed concentrates are clean, mold-free commercial

Body Score Your Lamas

A fat llama or alpaca is not a healthy llama or alpaca. This is especially true in hot, humid climates where heat stress kills. Learn to assess body condition, keeping in mind that in any herd there are a few individuals that are leaner or chunkier than the norm.

Unless you own lama-size livestock scales (and most of us don't), it's important to learn to body score your llamas and alpacas. Body scoring is the art of establishing comparative weight by feeling an animal in various locations and comparing your findings against a weight-scoring chart.

You'll find excellent llama body-scoring information at the Southeast Llama Rescue Web site (see Resources); at the site, click on *Llearn Llama*, then *Body Scoring*. An unusually fine alpaca body-scoring download is available for free at the Alpaca Association New Zealand Web site (see Resources); at the site, click on *Alpaca Health*, then on *Body Condition Score [BCS] of Alpacas)*.

High-quality commercial concentrate mixes take much of the guesswork out of feeding.

mixes (major companies such as Mazuri, Agway, Blue Seal Feeds, Buckeye Feeds, and Dynamite make them) or a grain mixture formulated specifically for your animals' needs. Be sure to store it where your llamas or alpacas can't break in and eat their fill, and where birds, cats, and wildlife won't contaminate it with their droppings.

Minerals

Always provide a high-quality, loose mineral mix formulated for your type of lamas and your locale. Place it where other species sharing their living space won't poop in it. Don't use mixes formulated for other species because they often contain copper in quantities that could be toxic to your llamas and alpacas. (The exception might be no-copper sheep minerals, but lama-specific products should be your first choice.) If your lamas cohabitate with another species, it's important that you place the other species' minerals where your lamas can't reach them.

Cool, Clean Water

The cheapest, most essential nutrient of all is water. The average llama or alpaca downs 5 to 8 percent of its

Lamas need access to clean drinking water at all times. This simple automatic watering setup is ideal for a few head of llamas or alpacas.

body weight in water every day, and it requires 10 to 18 percent of its body weight in hot weather or when lactating. Lamas can't thrive without 24/7 access, all year round, to clean, good-tasting water. They need it to maintain their digestive health; some males and geldings form urinary calculi unless they drink enough of it; and lactating females require water to make milk.

Don't skimp. Keep those tanks and buckets filled and clean. Consider installing automatic watering fixtures, but if you use them, clean and check them on a daily basis to make certain your lamas have drinking water on demand. (See chapter five for more information on running water and troughs.)

Hey, Hay!

As noted above, hay is critical to a healthy lama's diet, so let's consider it in greater detail. When selecting hay, keep in mind that quality is more important than type. Properly baled coastal Bermuda, for instance, is higher in protein and other digestible nutrients than is rained-on first-cutting alfalfa. A note about fescue hay, however: much of the fescue grass grown in the United States and Canada is infected with an entophytic fungus that produces several types of alkaloids known to be toxic to animals, especially during pregnancy. Although the jury is still out on whether lamas are susceptible to fescue toxicity, if another type of quality hay is available, it would be best to avoid fescue until the jury comes in.

Hay should smell fresh, never sour or musty. It should be 100 percent mold and dust free. Reject new bales that seem unusually heavy for their size or that feel warm to the touch; they weren't fully dry when baled. Dampness generates heat, which in turn causes the hay to mold or, worse, triggers spontaneous combustion.

The only way to gauge your hay's nutritional value is to buy tested hay or test the hay yourself. Your county extension agent has the tools and know-how to help you do it correctly.

Good Hay Versus Bad Hay

Characteristics	Good Hay	Bad Hay
Color	Alfalfa should be dark green unless it was treated with propionic acid, then it will be neon green. Good grass hay is light to medium green.	Yellow hay was dried too long in the field; it lacks the nutrition of greener hay. Sometimes one side of a bale is yellow from sun bleaching in storage; if the rest of the bale is green, it's OK. Light brown hay was damp when baled. It smells musty, and flakes tend to stick together; mold may be present. The bale is stiff, and the strings lack elasticity. Don't feed it. Dark brown hay was rained on; inside, it's generally caked with mold. Don't feed it for any reason!
Texture	Stems are easy to bend, relatively short, and slender. Bales easily separate into individual flakes.	Stems are thicker and often woody, and they crack easily when bent. Bales are difficult to split into flakes.
Leafiness	Roughly 60 percent of hay's total digestible nutrients (TDN), 70 percent of its protein, and 90 percent of its vitamin content are in its leaves.	Poor hay is mostly thick stems.
Weight	Good baled hay is fairly lightweight and easy to lift in proportion to the size of the bale.	Bales that are unusually heavy in proportion to their size are generally moldy, or they contain rocks, dirt, saplings, or other foreign matter.

Characteristics	Good Hay	Bad Hay
Mold	Good hay is *never* dusty or moldy.	If hay smells sour or musty, the center of the bale seems matted together, or hay is hard to separate into flakes, it's moldy.
Weeds	Good hay contains few, if any, weeds.	Bad hay is weedy, ranging from a few flowers to whole saplings; it's never a good buy.
Infestation	Good hay is hay and nothing more. Almost anything can be baled into poor quality hay, from feed sacks to barbed wire to dead cats.	*Always* discard hay containing insects and animal parts, no matter how small they are; they can carry deadly diseases—better to be safe than sorry.

Once you've purchased quality hay, keep it that way! Store it up off the ground to prevent spoilage, and preferably keep it covered so that birds, cats, and the like don't soil it.

Ask the seller to open several bales so you can properly evaluate the hay. Most types of hay should be green inside. Hay that is brown or yellow inside has been rained on or sun bleached in the field before baling. In both cases, the nutritional quality will be much lower than that of properly put-up hay. However, don't worry about slight discoloration on the outside of the bale, especially in hay that was stacked in the sun.

Neon green alfalfa hay has been treated with a mold-proofing preservative called propionic acid; it's a naturally occurring organic acid that acts as a fungicide. According to the Ontario Ministry of Agriculture, Food, and Rural Affairs bulletin *Preventing Mouldy Hay Using Propionic Acid:* "Hay treated with buffered propionic and other organic acid products is safe to feed to livestock. Propionic and acetic acids are organic acids that are produced by microbes in the rumen (and the cecum and colon of horses) and then used by the animal as part of the digestion process."

To minimize metabolic disease caused by changes in hay type or quality, buy as much hay at a time as you can properly store. If you can, buy tested hay. It's the only way to know for sure how much nutrition you're getting.

Shop around; you'll often find top-quality and very poor hay selling at the same prices. Don't depend on word of mouth, the local sales barn, or the classifieds when buying hay. For a wider choice of sellers, pick up a list of hay suppliers at your county extension office or peruse hay lists online at the U.S. Department of Agriculture Farm Service Agency's Hay Net Web site (see Resources).

Come to a mutual understanding with hay sellers before you finalize your purchase: iron out the specifics about price, delivery terms, whether the seller will hold additional hay for you, who is responsible for hay that doesn't meet your standards, and so on. Stack baled hay properly, up off the ground on wooden pallets or tires, under cover, where birds and other animals are unlikely to soil it.

Big Bales—or Not?

In some parts of the United States, it's becoming increasingly difficult to buy standard rectangular bales of hay weighing 45 to 100 pounds. It's cheaper, faster, and far easier for producers to put up hay in big round bales. These range in size from about 400 to 1,200 pounds. Big round bales stored under cover (and not stored directly on the floor or ground) can be fed to llamas and alpacas but with several caveats.

Make certain a bale is good *all the way through* before feeding it to livestock of any kind. Big bales are rolled up in layers, and it's not unusual to find moldy layers deep inside the bale. Remove the bales' strings; they're strong, long, and dangerous. Hay strings and livestock don't mix.

Livestock of every sort waste an incredible amount of hay unless big bales are fed from a bale feeder, which aren't necessarily safe. If you have such a feeder, check it at least twice a week for broken welds and other damage that your llamas and alpacas could injure themselves on. Remove the feeder if crias set up camp inside the ring.

If you don't have enough lamas to eat a big bale in three or four days, don't use the bale in an outdoor location; exposure to the elements quickly causes hay to start to go bad. Instead, feed the bale indoors or store the bale someplace under cover where your animals can't reach it, on a wooden pallet or on old tires to get it up off the ground. Unwind it as you need it, feeding just enough per day so that your lamas eat all or most of the hay you give them.

Large bales kept undercover are a better buy than field-stored bales, which in areas of heavy rain- or snowfall can represent up to 35 percent waste. If stored outside, they should be placed on pallets or poles to get them up off the ground and covered with plastic tarps only after they're fully cured (covering too soon or when they're damp from dew or being rained on causes excess spoilage). Don't store them close together, abutting each other from the sides (rather store them in a line, end to end), and don't stack them; both practices trap moisture between the bales and ultimately ruin the hay.

When good hay (like the quality bale of Bermuda hay on the right) isn't available, bagged grass or alfalfa hay works well.

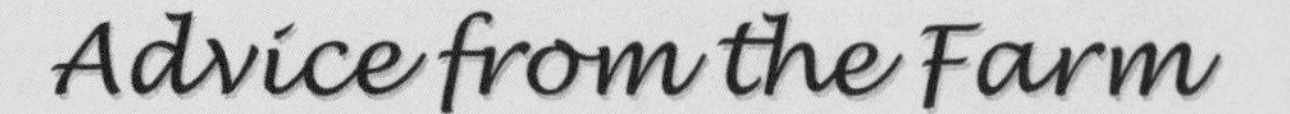

Advice from the Farm

The Best Ways to Treat a Lama

The experts offer their thoughts about lama treats.

Give Them a Hand

"Some people discourage hand-feeding, but we hand-feed—always have, with no problems. Our guys love peanuts in the shell. They're also big on carrot pieces, Cheerios, Triscuits, and thinly sliced raw sweet potatoes or apples."

—Deb Logan

Watch Out for the Gas

"Alex enjoys all kinds of stuff from our garden and orchard—his sheep do, too. Just don't give them a whole lot at one time, and always chop things into smallish pieces. One of my sheep once choked when he tried to swallow an apple whole; think what something like that would do inside Alex's long neck! And don't give ruminants anything from the onion or cabbage families because the first are toxic and the second causes (need I tell you) excess gas."

—Jan Johnson

Mix It Up

"I developed these bars while using click and reward training to teach my llama, Betsy, to load into a minivan. I never used them as a treat, but exclusively as a reward. I tried to fill the bars with good things that a llama would like. Other than all the salt, I like the llama bars, too. Just dump all the ingredients into a large mixing bowl, then mix with a mixer.

2 cups uncooked oatmeal (I use Old Fashioned oats)
2 cups whole wheat flour (I use coarse stone ground)
2 cups shredded carrots
1 tablespoon salt (too salty for me, but Betsy likes it)
½ cup vegetable oil (I use extra virgin olive oil)
1 cup molasses
½ teaspoon baking soda sprinkled on top of the other ingredients (the baking soda may need to be adjusted for a lower elevation)

"Mix until all ingredients are evenly distributed in the mix. Lightly oil a 9 x 13 inch cake pan. Press the mix into the cake pan. Bake at 350 degrees until an inserted toothpick comes out clean, or test with a light finger pressure as you would brownies. At this elevation it takes about 45 minutes, but everything bakes differently at 7,700 feet.

"Betsy and her friend Muffin periodically change their taste for rewards. A reward that has been working well may suddenly stop working. Also, they watch each other; thus, when one stops accepting a particular reward, the other one will soon stop also."

—Bob Huss

When You Can't Find Good Baled Hay

We live in an area often hard hit by drought. Sometimes quality hay trucked in from other states is available, sometimes it isn't. In a pinch, we've learned it's easier (and better) to feed alternatives when even nasty hay brings top-dollar prices.

When the going gets rough, pick up the phone and call the feed stores in your area. Chances are, they carry one or more types of bagged hay. Most brands of bagged hay are packaged in 40-pound, plastic-wrapped squares of compressed, chopped hay. If you feed your llamas or alpacas from outdoor feed bunks, prevent waste (especially on windy days) by pouring just enough water over the hay to dampen it down; your animals are less likely to dribble it around.

Or try haylage. Haylage is a semi-wilted hay product (grass or alfalfa dried to 55–65 percent dry matter as compared to 82–85 percent in hay), compressed and sealed in tough plastic wrapping. Llamas and alpacas love grass haylage, and when handled correctly, it's first-class fodder. However, once bagged haylage is opened (or the wrapping is torn), mold spores begin to proliferate, so after three or four days, any uneaten feed should be discarded. In the United States, the major manufacturer of bagged haylage is Chaffhaye (see Resources).

Another option: feed hay pellets. They're readily available, and as most hay pellets contain adequate particle size necessary to maintain normal digestive health, they can be used to partially or totally replace baled hay. However, some llamas and alpacas tend to wolf down pellet feeds, and then they choke. So, if it's hay pellets or nothing, use them, but soak them for an hour or so before feeding to reduce them to a yummy mash. Or make it harder for greedy gobblers to hog down their pellets by placing fist-size rocks in their feed boxes that they'll have to eat around and thus slow down.

CHAPTER FIVE

Housing Llamas and Alpacas

Llamas and alpacas are among the easiest livestock species to house and provide for, so you needn't spend a mint to add them to your hobby farm. However, they do require adequate shelter, fencing, and feed. Chapter four discusses feed. In this chapter, we'll address shelter and fencing needs.

Home Sweet (Lama) Home

In warm climates and on sunny days, natural shelter such as mature trees for shade and dense hedges or rock outcroppings for windbreak may be sufficient. However, in rainy and snowy climates, the animals also require man-made housing. And even in warm climates, they appreciate having access to housing. Shelters should be sound and roomy enough to accommodate every individual in the group. However, you don't need a lama mansion to house a few head of llamas or alpacas. Given a dry place to sleep in a draft-free shelter and a pasture or an outdoor exercise area to wander in when the mood strikes them, llamas and alpacas are quite content. They're the essence of simplicity to house and care for.

Basic Structures

Three-sided roofed structures called loafing sheds or field shelters, erected with their open sides facing away from prevailing winds, make ideal, inexpensive lama shelters. The rules of design are simple: allow at least 30 square feet of floor space for each llama and 20 square feet for each alpaca; make certain barn site drainage is adequate; and slope the roof away from the shelter's open side so rain and snow roll off the rear, rather than the front, of the structure.

This simple shelter provides weather protection for a few sheep and their guardian llama.

When building field shelters in cold northern climates, keep the roof height as low as possible, while allowing enough headroom for your animals to comfortably raise their heads. Low-slung roofs hold body heat at lama level. By the same token, high ceilings are better in southern climates, where heat stress is a serious summertime problem. (We'll talk about heat stress in chapter six.)

Structures such as commercially built or homemade, fabric-covered hoop houses or metal Quonset huts such as the ones manufactured by Port-a-Hut (see Resources) are also ideal for housing llamas and alpacas. Or you can house your lamas in an enclosed barn. However, before building a barn for your llamas or alpacas or renovating the barn you already have, visit your county extension office and ask for housing bulletins and plans specific to your climate and locale. Consider it your first stop for reliable, geographically correct building advice—and the services are free!

Bedding

Bed any sleeping and lounging areas with 4 to 6 inches of absorbent material, such as straw, discarded hay, wood shavings or sawdust, peanut hulls, ground corncobs, or sand. If you plan to harvest fiber, long-stem bedding such as hay or unchopped straw is better than shavings and other small stuff that works its way into your llamas' and alpacas' wool.

Because lamas create and use communal dung piles, even when housed indoors, you won't go through bedding the way you would if you were housing another livestock species. Simply remove excess droppings (don't take the whole pile; lamas expect you to leave their dung piles partially intact) and damp spots, then add just enough bedding to keep things dry. Clean everything out, back to floor level, several times each year and start again. This system, called deep litter bedding, is comfortable, warm, and simple to maintain.

Find a responsible way to dispose of used bedding. Compost it or give it away, but don't let it pile up, especially on nonagricultural, zoned properties.

Points to Keep in Mind

However you ultimately choose to house your lamas, keep the following points in mind. Lamas like to see other lamas. Design barn interiors so that no individual feels isolated. Isolation leads to stress and stress kills. For the same reason, provide safe getaways where stressed, low-ranking llamas and alpacas can escape bossier herd mates. It's best to have at least one enclosed stall to use when you have a sick lama or a female with a newborn cria. If you have a radiant barn heater, put it in this stall. If other animals, particularly horses or ponies, are part of the herd dynamics, provide at least two exits so cranky herd members can't corner a passive lama and beat it up.

Llamas and alpacas kept in poorly ventilated, tightly enclosed winter housing often suffer from deadly respiratory ailments such as pneumonia. Don't shut all the doors and windows—healthy llamas and alpacas can handle considerable cold. If a lama gets sick or crias are born during a cold snap, blanket them or install a livestock-approved radiant heater in a single stall instead of heating the whole barn. If you absolutely must use a heat lamp, make certain it's securely tied (don't suspend it by its cord), and hang it high enough to prevent bedding from igniting from the heat or curious mama lamas from pulling it down.

Packed dirt or stone floors are easier on lamas' legs and feet than concrete. If the building you refurbish for lamas has concrete floors, top standing areas

Loafing sheds provide a comfortable resting area away from the hot sun and inclement weather.

Advice from the Farm

Roofing and Fencing Them In

The experts talk about housing and fencing for lamas.

Escaping the Great Outdoors

"Some say shelters should be optional and that their animals will not use them. From my experience, both llamas and alpacas will use a shelter when they think they can easily escape if they have to. We have a small barn with an 8-foot door and a 20-foot-deep 'porch,' which is equipped with fans that are operational in the summer months. Our llamas always take shelter in moderate to heavy rain as well as during the heat of the day, and I believe they are healthier for it."

—Deb Logan

Sleeping with the Sheep

"The first year we had Alex, I was afraid he'd get cold during the long weeks of really cold winter. But he and the sheep preferred to stay outdoors except when it was really windy and during all but the very coldest nights. We use submersible heaters in our water troughs and make sure there is free-choice grass hay and dry bedding in their shelter. That way they can choose to sleep in or out, and we sleep easier knowing they have that choice."

—Jan Johnson

With All Due Respect

"Llamas and alpacas are traditionally very respectful of fences. Fences are more to keep other animals out rather than to keep the camelids in.

"We recommend a minimum of 4-foot fencing, although we like 5-foot even better. You can use 4-foot stock fencing and just add a strand of wire on top to add height. Welded wire is good if you have uneven terrain; 'no climb' horse fencing is great but is stiffer and more difficult to keep close to the ground unless the ground is perfectly flat.

"Llamas and alpacas are naturally insulated by their wool, so electric fences don't impress them much. If you use an electric fence, make sure the strands are very visible. Llamas respect a visible barrier and hopefully anything else attempting to get in will get zapped.

"Barbed wire is a bad thing. A llama or an alpaca that gets tangled in barbed wire will typically kush rather than panic, as a horse would, but they can get deeply gouged. Animals can also get scraped about the face and ears if they decide to stick their heads through to munch on whatever is on the other side."

—Deb Logan

with rubber stall mats made for horses. Wood floors are slippery when wet, and they tend to rot.

Lamas are exceedingly curious creatures. They'll thread their fragile necks through the darndest places to get a better view, and when they do, they sometimes get injured or stuck. Any type of fencing material with 6-inch or larger squares (4-inch or larger squares when fencing for alpacas) should never be used around lamas. Especially avoid the use of cattle panels with large openings when you are building accommodations for lamas—strong steel rod is next to impossible to cut through in an emergency. In addition, protect glass windows and electrical wiring with screens or conduit or keep them well out of the reach of lamas.

Intact male llamas are stronger than you probably think, especially when challenging another male or enamored of a nearby lady friend. Build their shelters, pens, and fences using stout, sturdy materials, and try not to place two intact males side by side.

Dining Accommodations

No home is complete—or inhabitants happy—without accommodations for food and drink. You'll need hayracks, grain feeders, and mineral feeders as well as water troughs and buckets.

Places to Eat

Build or buy sturdy, safe hayracks and grain feeders for your lamas. Lamas fed on the ground are prone to internal parasite infestation, and they just plain waste a lot of feed. Feeders should be designed so the animals can reach their feed but not get their heads stuck while eating. If you plan to raise crias, hay feeders must also be cria-proof lest they crawl inside and get stuck.

Creep feeders provide a place for crias to eat supplementary concentrates without allowing their dams to pig out.

Don't store feed where lamas can help themselves. Overeating, especially of grain or rich legume hay, can cause enterotoxemia and bloating; both conditions are fairly uncommon but often fatal. Store grain in lama-proof covered containers with snug lids. (Fifty-five-gallon food-grade plastic or metal drums and decommissioned chest freezers all work well.) Secure the feed room door with a lama-proof lock; standard, simple closures are child's play to a nimble-lipped llama or alpaca.

Always provide access to lama-appropriate, free-choice minerals, including salt, in a mineral feeder or open pan. Make sure new lamas know where they are.

Places to Drink

Lamas require lots of fresh, clean water, kept cool in the summertime and thawed when temperatures dip below freezing.

A simple, heated trough provides necessary drinking water when the mercury falls below zero.

Install running water and electricity (you'll need electricity for handling nighttime emergencies, too) to your barn or shelter. Barring that, locate the structure within easy garden hose and extension cord reach of existing utilities.

If you have more than just a few lamas, provide multiple watering troughs or buckets. It's easier to dump, scrub, and disinfect several smaller containers than it is one huge trough, and having multiple water sources prevents bossy, high-ranking lamas from guarding it so others can't drink.

Pasture Perfect

Lamas love their freedom, and they thrive on grass. Whenever possible, provide your llamas and alpacas with a place to loaf and graze instead of keeping them in a barn lot or corral. Keep in mind that all pastures are not created equal; a small, well-cared-for meadow of fertilized native grass is better than a huge area overgrown with thistles and weeds. By the same token, lush spring pastures and meadows of legume hay are too rich for a llama or alpaca's metabolism. If you use them, grazing time must be carefully monitored lest lamas become too fat. (See chapter four for more information on pastures, including what plants must be avoided at all costs.)

Llamas and alpacas thrive on pasture, but they can't simply be turned out and forgotten about. They'll need suitably sized, safely and securely fenced pastures; ready access to shelter; a copious supply of clean water; supplements such as salt, vitamins, and minerals (and sometimes additional supplements such as grain); protection from biting flies and other insect pests; companionship; and daily monitoring, along with routine worming, toenail, and coat care.

Fencing for Your Lamas

The cardinal rule of lama keeping, especially if you plan to pasture your llamas or alpacas, is don't buy them until after you've installed fencing strong enough and tall enough to keep them in and predators out. Start with a large pen or a small paddock if you like (a rule of thumb: one acre of land will carry three to five llamas or five to eight alpacas) but make it lama-proof and highly predator-resistant!

Llamas and alpacas (especially crias) can squeeze through incredibly small gaps. I recently watched our yearling llama kush, roll onto his side, and then inch his way under the bottom rail of a round pen panel with its bottom rail just 14 inches off the ground. Standard plank and post fences won't contain Houdini-minded llamas, and alpacas and crias will definitely crawl under them unless the planks are set very close together or the fence is lined with woven wire fencing.

Check It Out

Check on your lamas at least twice every day (more often is obviously better). Inspect them for illness and injuries, and remove major debris such as sticks from their fiber. Check the water supply; make certain automatic waterers are functioning and water receptacles are filled and tidy. Do the same for loose mineral feeders. Make certain electric fence chargers are working, and make a quick sweep of shelters, inside and out. Mist with premise insecticide and remove wasp nests.

Carefully inspect all shelters on a weekly basis, combing them for hazards and mending any you find. Strip soiled bedding and replace it with fresh. Walk the fence lines and pasture, keeping an eye out for hazards such as loose wire and watching for poisonous plants.

Sis takes her week-old cria, Kamilla, out to the wooded pasture to browse. Wooded pasture serves double duty, providing forage and light shelter.

You have some choices in the material you use for your fences. I recommend that you use woven wire fencing or electric fencing. Although barbed wire is the classic stockman's fence, *don't use it.* Animals and barbed wire don't mix. Llamas and alpacas are less likely to injure themselves on barbed wire fence than most livestock species are, but it can happen. If you already have barbed wire on your farm, especially if it was used to build cross fences that running lamas could crash into, rip it out and replace it with something friendlier.

High-tensile smooth fencing, electrified or plain, was once the livestock fence of choice, but too many serious injuries occurred when animals hit tautly stretched wire while traveling at a high rate of speed. Since lamas are less likely to gallop blindly into fences than many other species are, high-tensile wire fences might be an option on your farm, but not a good one.

Lamas put their heads in the darndest places! To avoid injuries and damage to fleeces, try to avoid fencing through which curious lamas can weave their long necks.

Whatever you choose to use, walk every fence line, making repairs as needed, before turning llamas or alpacas out to pasture. Replace unfriendly materials such as strands of saggy high-tensile fencing or barbed wire. If fences rely on electric current, make certain the fence charger (these are also sometimes called fencers or energizers) works and the fence is hot.

Woven Wire Fencing

Correctly installed woven wire (also called field fence or field mesh) is the most secure form of affordable lama fencing, making it ideal for perimeter or boundary fences and for using inside plank fences to render them lama tight. Four- or five-foot-high woven wire will contain most lamas; installing one or two strands of electric wire (offset to the outside) above woven wire will help keep predators away from your herd.

Woven wire fencing is fabricated of smooth, horizontal wires held apart by vertical wires called stays. It's sold in galvanized, high-tensile, and colored polymer-coated high-tensile versions. (High-tensile woven wire costs more than standard woven wire fencing, but it is rust resistant, sags less, and is lighter in weight.) All of them have verticals placed at intervals ranging from 6 to 12 inches and come in heights ranging from 26 to 52 inches. Horizontal wire spacing generally increases as the fence gets taller. It is important to avoid woven wire with openings more than 6 inches square (4 inches square if your animals are alpacas); curious llamas and alpacas constantly poke their heads through inviting openings, damaging valuable fiber, injuring their eyes, or even getting stuck—all situations you definitely want to avoid.

When buying woven wire, check the numbers: 8/32/9 fencing has eight horizontal wires, it's thirty-two inches tall, and it has vertical stays every nine inches. Woven wire is sold in twenty-rod (330-foot) rolls. Wood or steel posts erected at 14- to 16-foot intervals are needed to support it. Disadvantages of woven wire are its cost and the time and effort required to install it. Advantages: it's safe and looks good, and once it's properly installed, it requires very little upkeep.

High-tensile (also called New Zealand) fencing can work well for lamas but should never be used where equines are kept.

Electrified Fencing

Electrified tape fences are effective and relatively inexpensive to build and maintain. These fences usually utilize step-in plastic posts, so they're the essence of simplicity to move. Portable fencing is ideal for subdividing pastures for rotational grazing; erecting small pastures for special needs animals such as females with new crias or animals in quarantine; fencing steep, rocky, or otherwise uneven terrain; and fencing rented land.

If you live on hard or rocky terrain, however, using step-in posts is probably not an option; you'll have to opt for standard metal posts. Electrified tape comes in an array of widths and colors; choose a product wide enough for llamas and alpacas to see.

Keep in mind that any electric fence can fail if the fence charger that powers it isn't up to the job. Fence chargers are sold according to the voltage and the number of joules they put out. (A joule is the amount of energy released with each pulse.) A single joule will power 6 miles of single-wire fencing; a 4½-joule fencer will energize 20 to 60 acres, depending on the length of the fence and the number of wires that are used in the fence's construction.

Electric fences must be properly grounded, however, or they lose that vital zap. Yet according to various university studies, an estimated 80 percent of the electric fences in the United States are improperly grounded. At least three 6- to 8-foot-long ground rods should be used with each fence charger. Carefully follow the fence charger manufacturer's instructions when putting them in.

You must train your lamas to respect electric fences because electrified fences are psychological, rather than physical, barriers. A good way to do this is to suspend tops cut out of metal food cans from the fencing, using loops of conductive wire. Place two or three curious llamas or alpacas in a good-size electrified enclosure and let nature take its course. Once they've shocked their noses, most lamas tend to honor electric fences.

Steps to Better Electric Fences

1. Don't skimp on wire or tape. The larger the wire and the wider the tape, the more electricity each can carry.
2. Don't space strands of tape or wire too closely. To get the most from your fence charger, keep strands at least five to seven inches apart.
3. Use quality insulators. Sunlight degrades plastic; choose high-quality insulators, preferably a brand treated to resist damage done by ultraviolet (UV) light.
4. Think big. The more powerful your fence charger, the fewer problems you'll have, so pick a model that packs a punch.
5. Follow the instructions. Mount solar fencers directly facing the sun; install ground rods as directed in the user's manual. Unless you install your fencer charger correctly, it can't do the job for which it was designed.
6. Buy a voltmeter. A good one costs fifty to seventy-five dollars, and it'll help you keep your fences in the pink. Check voltage every day; if a fence runs low on voltage or shorts out, you want to know it and correct the problem right away.

Fence Posts

Where rocky soil isn't a problem, flexible plastic step-in fence posts work well with electric tape or wire fencing. Otherwise, opt for wooden posts or steel "T-posts." Steel fence posts lack the eye appeal of wooden ones, but T-posts are fireproof, long-wearing, less costly, lighter in weight, and easier to drive than wooden posts are. Larger animals such as horses and llamas have been known to impale themselves on the tops of steel posts, so it's best to cap them with plastic T-post toppers if you use them. Expect unbent T-posts to last twenty-five to thirty years.

Wooden posts come in treated and untreated varieties. Treated posts last twenty to thirty years; untreated ones, from two to twenty years depending on the type of tree they're made from.

Primary pasture gates should be wide enough to drive a truck or a tractor through them; 10 feet wide will usually do, but a 12-foot-wide gate is better. For convenience, add walk-through gates in all high traffic areas.

Catch Pens and Restraint Chutes

Do you really need a catch pen and a chute? In a word, yes—unless you can walk up to every one of your lamas and slip a halter on it, then give it a shot, shear it, and trim its toenails without a huge fuss.

An easy-to-construct catch pen consists of four chained-together, 10-foot-long, 5-foot-high round- or square-tube livestock or round pen panels (buy them at tack and farm stores)—8-foot-

While commercial squeeze chutes are obviously nicer, a simple, two-gate squeeze chute like this one will also do the job.

long, 4½-foot-high panels work well for alpaca catch pens. Set the pen up in your pasture, then open the gate to let your curious lamas go inside and find a yummy treat you left in there. Once they associate the pen with something pleasant, it's easy to use herding poles to guide them into the catch pen when you want to handle them.

A restraint chute can be as fancy (and expensive) as ready-made handling equipment designed for llamas or alpacas or as simple as a pipe gate hinged to the side of a sturdy building. If you'd like to build something in between, download llama chute plans at the Southeast Llama Rescue Web site (see Resources); click on *Llearn Llama*, then *Build a Restraint Chute*. Free alpaca chute plans are downloadable from the Southern Iowa Alpacas Web site (see Resources); click on *Alpacas Magazine Husbandry Hints Articles*, then scroll down to the Spring 2004 issue. These instructions are good ones—llama owners may want to read this download, too.

A suitable catch pen is a lama-keeping essential. This one is made of round pen panels and is the perfect height for tall llamas like Bandit.

Predator Protection

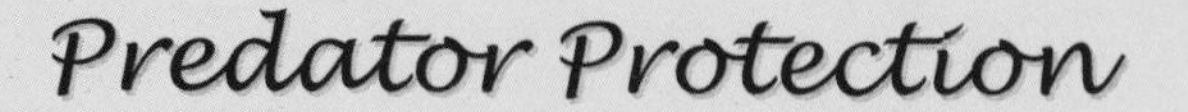

Although the U.S. Department of Agriculture's National Agriculture Statistics Service (NASS) doesn't track the number of alpacas and llamas that die each year, it does keep track of how many U.S. sheep and goats die and the causes of death. The agency periodically publishes findings in a report titled *Sheep and Goats Death Loss*. According to the May 6, 2005, edition, predators killed 155,000 goats during 2004, accounting for slightly more than 37 percent of all goats that died in the United States that year. Sheep figures are just as astounding: predators killed 225,000 sheep during 2004 (also totaling 37 percent of total losses). Small ruminants (sheep and goats combined) were killed mainly by coyotes (more than 60 percent of the total) but also by dogs, mountain lions, bears, foxes, eagles, bobcats, and other species (such as wolves, ravens, and black vultures). Keeping in mind that alpacas, adults and crias alike, are the same size or smaller than most goats and sheep, you'll agree that predation is a potentially serious problem, even in relatively populated areas, where free-roaming dog predation poses a major risk.

To protect your lamas, you should:

- Install predator-proof fences.
- Pen llamas and alpacas (particularly females with crias) inside very secure fencing, close to your house at night and whenever no one is home.
- Add guardian animals to your herd dynamics.
- Preferably, do all three of the above.

Llamas as Guardians

In certain situations, llamas themselves make outstanding herd guardians, but only in low-risk locales. A llama (or even several llamas) can't effectively repel aggressive packs of dogs or coyotes, much less the big predators, such as mountain lions and bears. In high-risk situations, guardian llamas are often maimed or killed while attempting to protect their charges.

Guard llama and cria.

Donkeys as Guardians

Some farmers prefer donkey guardians. Donkeys require no specialized training. And, they instinctively dislike the canine clan, so most will attack dogs and coyotes tooth and hoof. Since donkeys have keen hearing and good eyesight, dogs and coyotes rarely sneak past a donkey standing guard.

Guardian donkeys must be standard "burro" size or larger; miniature donkeys require guardians of their own. Jennies (females) work best. Geldings work well, too, but never keep a jack (an intact male) with your llamas or alpacas, as they can be aggressive toward herd mates they dislike and have been known to kill newborn crias. And not all donkeys are interested in bonding with another species, especially when other equines are within sight and smell.

That said, in one survey (reported in the Colorado State University publication *Livestock Guard Dogs, Llamas and Donkeys*), 59 percent of Texas producers who use guardian donkeys rated them good or fair for deterring coyote predation and another 20 percent, excellent or good; and 9 percent of the sheep and goat producers polled for the 2004 National Agriculture Statistics Service survey successfully keep guardian donkeys, too.

Dogs as Guardians

For many thousands of years, European, Middle Eastern, and Asian guard dogs of dozens of types and breeds have protected herds of goats and flocks of sheep from predation by wolves, bears, jackals, and human thieves. Eventually, some of those breeds immigrated to North America.

According to National Agriculture Statistics Service figures, nearly 32 percent of American sheep and goat producers use livestock guardian dogs. Because they've been bred to be guardians for thousands of years, when bonded with a herd from puppyhood on, livestock guardian dogs require little or no specialized training. Once bonded with the animals to be protected, a guardian dog willingly stays with them and fearlessly protects them twenty-four hours a day. Several major livestock guardian breeds, especially Great Pyrenees and Anatolian Shepherds, are readily available throughout North America at reasonable prices.

Half Anatolian Shepherd, half Great Pyrenees.

Great Pyrenees guard dog.

There is one major drawback: livestock guardian dogs tend to catnap through the day and bark throughout much of the night, because that's when predators are most active. This effectively warns most predators away but sometimes causes problems with light sleepers and unsympathetic nearby neighbors.

In addition, one dog can't effectively protect livestock from attack by large packs of dogs or coyotes nor from predators such as mountain lions and bears. Where heavy-duty predators are the norm, a pair or trio of dogs works best—one to herd your lamas to safety while the others deal with the invaders themselves.

A word to prospective purchasers: buy from responsible livestock guardian dog breeders. Show dogs have been selected for beauty and gait, not guardian ability. Stick to proven adults or buy puppies raised with livestock by working guardian moms.

CHAPTER SIX

Llamas and Alpacas in Sickness and in Health

The trick to having healthy, happy llamas and alpacas and to protecting them from needless injuries is to start with healthy stock and keep it that way. You should begin by learning all the signs that tell you whether a lama is healthy or sick—and by not bringing any unhealthy ones home. Then you'll need to know what daily, weekly, monthly, and yearly tasks you should perform to keep your lama doing well. Finally, you should be able to recognize and be on the lookout for signs of illness and know what kind of care to give. And, of course, having a good lama-knowledgeable veterinarian from the beginning is crucial.

Keep Your Lamas in the Pink

It's easy to keep your llamas and alpacas healthy if you nip potential problems in the bud. Quarantine all incoming animals—make no exceptions—and disinfect your quarantine pen after each occupation. Maintain a hospital pen for sick animals (use your quarantine pen if it isn't already occupied); don't leave them with the rest of the herd.

Monitor your llamas and alpacas at least twice a day (the more often you do it, the better). Count noses and make certain all animals appear to be healthy and uninjured (refer to the Healthy or Sick? chart in chapter two). Address sickness or injuries immediately; don't wait to see whether an animal gets better without treatment. If you don't know what's wrong with your llama or alpaca or you're not positive you know how to treat what ails it, always call your vet. Closely monitor pregnant females; be there when crias are born. Know how to assist if you have to, and keep a well-stocked birthing kit on hand.

Did You Know?

- The air on the high Andean Altiplano is very thin. One way lamas have adapted to this environment is that their red blood cells are oval instead of round; this helps them take in more oxygen.
- Llamas are frequently used by physiologists to study hypoxic (oxygen saturation) stress because of their ability to live successfully at high altitudes.

As mentioned in chapter four, prevent serious digestive illnesses such as enterotoxemia and bloat by making dietary changes over a period of time (a week to ten days works well). This allows your animals' gut microbes time to get used to new types or quantities of feed. Feed quality hay; don't feed moldy bales and avoid dust.

Assemble a first-class first aid kit and learn how to use it. Keep the kit where you can find it in an emergency. As you use an item, replace it. Do an inventory every six months; check expiration dates, and discard outdated products. Vaccinate for enterotoxemia and tetanus no matter where you live. Discuss a vaccination program with your veterinarian, and add other immunizations recommended for llamas and alpacas raised in your locale.

Maintain your llamas and alpacas in clean, dry surroundings. Provide adequate drainage in barn lots and shelter areas, and make sure your animals have adequate, draft-free shelter. Police your pastures, pens, stalls, catch pens, and restraint chute, and remove protruding nails, exposed wire, sharp edges on metal buildings, hornets' nests, and so on. Set up a weekly or monthly schedule and follow it.

To prevent predator-related deaths and injuries, provide adequate predator protection (see chapter five). Maintain good fencing, house your lamas close to human habitation from dusk through midmorning, and add livestock guardian animals to the mix.

Have dead animals, especially aborted fetuses, tested to determine cause of death; ask your veterinarian for details. Remove dead animals and birthing tissues immediately (don't allow livestock guardian dogs to eat them); if they aren't going to be necropsied or tested for pathogens, properly dispose of them by burning, burying, or composting.

Newborn crias are particularly susceptible to illness, so it's important to provide shelter from bad weather.

Trim toenails on a regular basis to maintain soundness and prevent nail deformities. Avoid stressing llamas and alpacas during handling, hauling, or weaning. Stress is the ultimate lama killer; avoid it however you can. Shear your lamas when they need it. A timely barrel cut goes a long way toward preventing heat stress.

Did You Know?

Most well-cared-for lamas live to be sixteen to eighteen years old, but many live into their early twenties. The world's oldest alpaca, a New Zealander called Vomiting Violet, died in 2005 at the age of twenty-nine.

Shear Happiness

Everyone who keeps lamas must know how to shear them, which runs the gamut from scissoring the fiber off barrels to crafting fancy 'dos. By shearing your lamas, you can harvest their valuable fiber and make them look pretty. But above all else, you must shear for your animals' health and comfort. Lamas are highly prone to heat stress (see the appendix, Lama Maladies at a Glance), and a heat-stressed lama can quickly die. I know because I lost my first, sweet llama that way.

Beautiful clips like the one this Klein Himmel Llamas llama is wearing are ideal but generally beyond the scope of first-time shearers.

To help prevent heat stress, shear every llama and alpaca every spring. In mild climates, judicious removal of fiber from the lama's undercarriage may be enough. (Lamas dissipate heat from their bellies, armpits, and groins, so it's important to free those areas of excess wool.) In colder regions, a full body clip or a barrel clip is much better.

Hairstyles for Llamas and Alpacas

The most basic hairdo is a full, all-over clip. To body clip, simply scissor, shear, or clip all of the fiber off of your lama save for that on its tail and, if you like the way it looks, on its head. This is a particularly popular 'do for alpacas.

Llama owners tend to prefer the blanket clip, in which all of the fiber from a point behind an animal's withers and elbows back to its hips and hind legs is removed.

Beyond these two basic styles, the sky is the limit. Examine the pictures in this book and visit farm Web sites listed in the Resources section for additional ideas.

Do-it-Yourself (or Not?)

If you can find an experienced lama shearer to do the job for you, you'll probably be money ahead—especially when saving quality fiber is an issue. If you choose to tackle it yourself, you'll need to buy (or borrow) the proper equipment.

Using self-opening scissors is an inexpensive way to shear one or two lamas. Fiskars scissors like the ones shown below are efficient and quick.

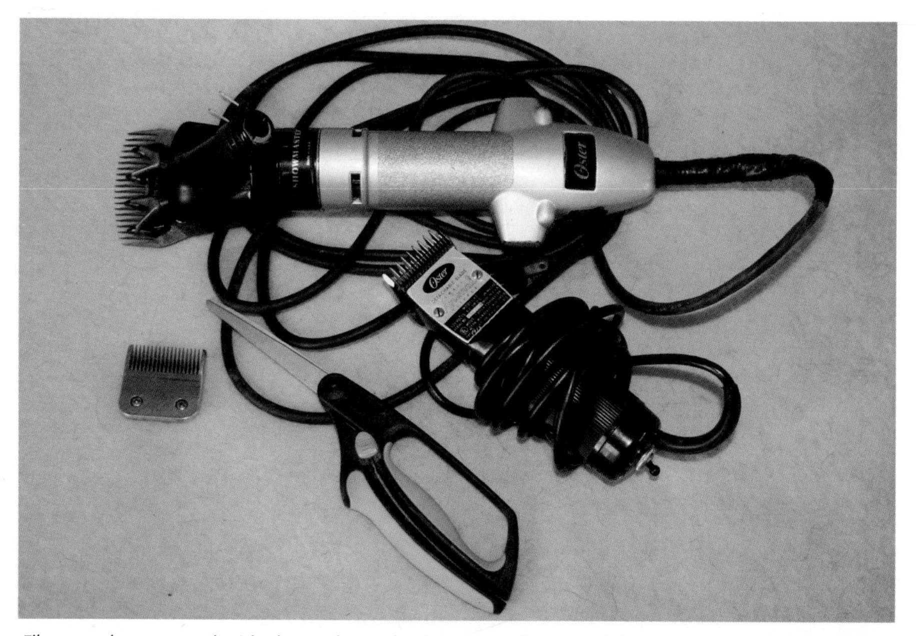

Fiber can be removed with sheep shears (top), clippers (lower right), or even comfortable scissors (lower left).

Secure your lama and away you go—no expensive, noisy equipment needed.

Hand shears of the type traditionally used to shear sheep work, too. They're faster because they easily cut a larger swath, but they're more cumbersome than scissors and you're more likely to cut your lama (a hazard of shearing no matter which tools are used). Most hand shears have five-inch blades, which can be hard to operate if you have small hands; if that's the case, ask for dagging shears with shorter, three-inch blades.

Most lama owners opt for electric shears or clippers. Shears have a head designed for shearing sheep and fiber goats; clippers have a head and blades designed for clipping horses, dogs, cattle, and the like. Either is suitable for shearing lamas.

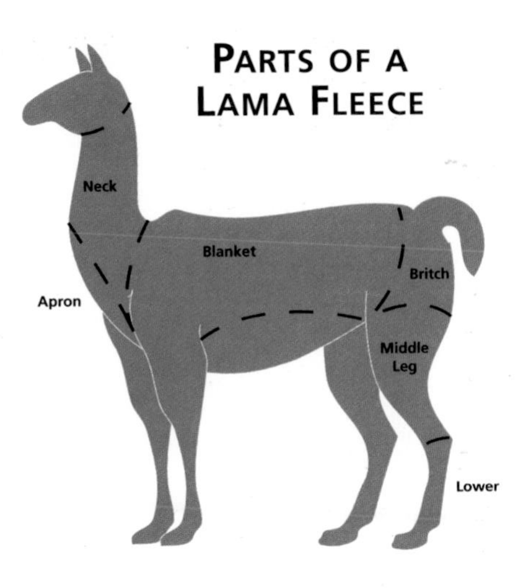

You'll also need a means to secure your lama while you work. Most lamas intensely dislike this process, so using a chute and a helper to stand near the lama's head to reassure it is a very good idea. Keep in mind that even easy-

People new to showing llamas sometimes think shearing will spoil their chances of placing. Not so! A great clip accentuates each individual's best points, such as this handsome guy's topline and hindquarters.

Here is Bandit, who is a short-wool ccara llama, before shearing. Unfortunately, his coat is very matted.

going lamas often spit at shearing time. I recommend wearing face shields or safety glasses.

Getting Down to Business

Before you begin clipping or shearing, use a blower (commercial units are available but inexpensive leaf blowers work well) to surface clean excess debris from the animals' fiber. Keep the blower nozzle far enough away that the blast of air causes fiber to fan out instead of tangling. Don't deep brush the lamas prior to shearing, and don't try to tease out stubborn mats using a brush. It hurts, and it doesn't work.

Have a plan. Take into consideration that pink-skinned lamas sheared to the skin will painfully sunburn, so don't close-cut unless you keep your lama inside. It's usually best to leave about an inch of fiber; this is where scissors and manual sheep shears shine because you can more easily manage the depth of cut.

It's best to start at the top of your lama's back and work down, making parallel cuts as you do.

As you work, constantly praise your lama, no matter how its behaving. Chances are it's badly frightened. Losing your temper will only increase its stress level and cause it to react worse the next

Before shearing, blow out as much debris as you possibly can, even when the lama's fiber is badly matted.

It's possible to shear with scissors but more time-consuming and less safe. We used them to remove Bandit's worst mats, but we switched to electric sheep shears to do the rest of the job.

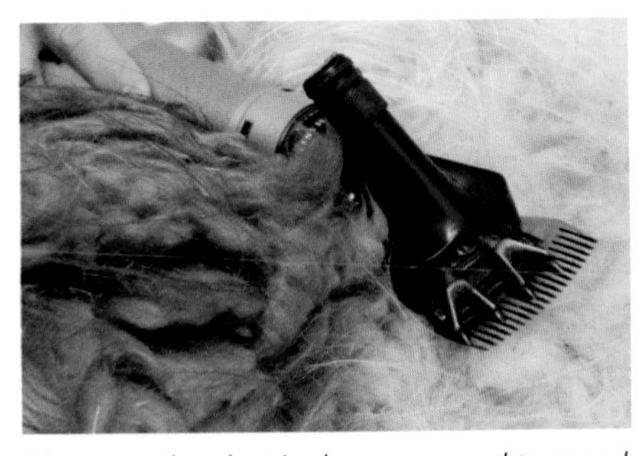

These are the electric shears we used to speed up the shearing process.

time you shear. Keep in mind also that lamas often react badly for the first ten or fifteen minutes then settle down nicely if you don't get mad. Grit your teeth, do your best, and watch out for flying spit!

If you're saving your lama's fiber, avoid the temptation to "even things up and make them look better" by making second cuts. Second cuts are short snippets of fiber produced by touching up here and snipping something there, and handspinners hate them. Instead, periodically remove shorn fiber (it tends to hang in sheets no matter which tools you use) to a clean, dry place. When you're finished removing first-cut, prime fiber, sweep the floor, *then* go back and even up the rest.

Don't be discouraged if the results aren't precisely what you hoped for. Mistakes grow out, and even the most pathetic hairdo looks fine in a month or two.

If you're saving fiber, save only the usable parts. Discard stubborn mats and dirty or badly sun-bleached fiber. Separate neck and leg wool from body fiber (the latter is the lama's best fiber), and place it in separate, large paper bags (clean grain bags work quite well) or old pillow cases. *Never* put it in plastic. Store it where marauding mice and moths won't spoil it.

For more insight into shearing lamas, run an Internet search using the words *shearing llamas* or *shearing alpacas*. It's a stressful job, but someone's got to do it, so for your lamas' sake, try to do it right!

We started at the back, lifting off the matted fiber as we sheared it loose.

If your llama is as badly matted as Bandit was, much of its fiber will be waste.

To Be (or Not to Be) Your Own Vet

It's often difficult to find lama-savvy veterinarians in parts of North America. Because of this, lama owners sometimes become their own vets. Maybe they shouldn't.

Serious illnesses sometimes resemble one another, and a life-threatening symptom such as scours (watery diarrhea) can be caused by several different conditions. Scours, for instance, could be a sign of coccidiosis, poisoning, or an internal parasite overload.

Most of the major pharmaceuticals used to treat llamas and alpacas are prescription drugs, all of which are off-label for lamas. You'll have to get the medications through a vet.

The most workable solution: find a veterinarian who's qualified to treat your animals and establish a working relationship (see Finding and Working with a Veterinarian in chapter two). At the same time, learn to address minor problems and routine veterinary procedures yourself. To do so, begin by putting together well-stocked first aid kits in your barn and in any vehicles you use to transport your llamas and alpacas.

Build a Better First Aid Kit

We pack our farm-based first aid items in two five-gallon plastic food service buckets fitted with snug lids. On the top and both sides we've affixed big Red Cross symbols using red duct tape, so the buckets are easy to spot when we need them. One is marked with SHEEP, GOATS,

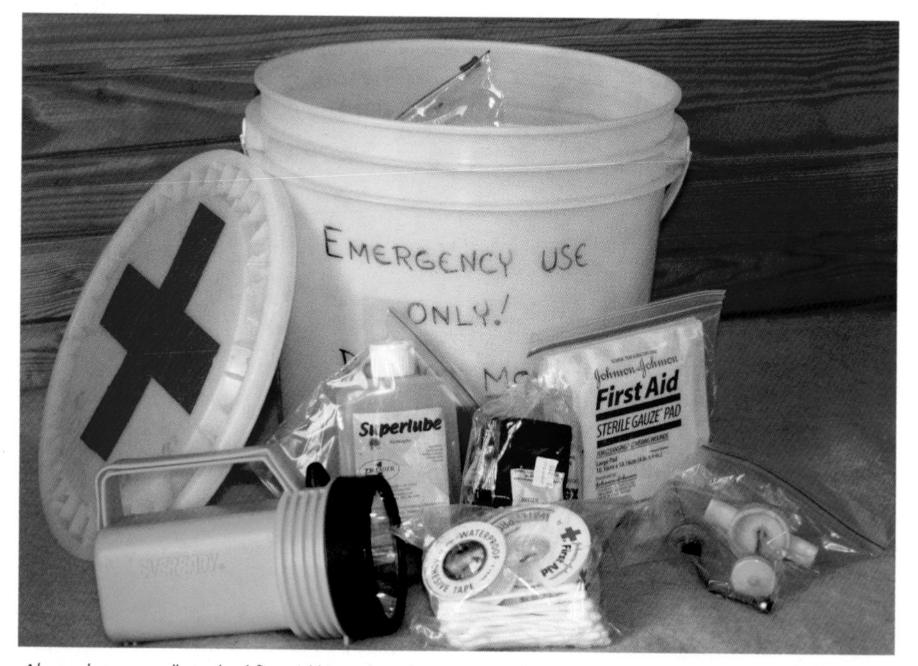

Always have a well-stocked first aid kit on hand—you never know when you or your lamas may need it!

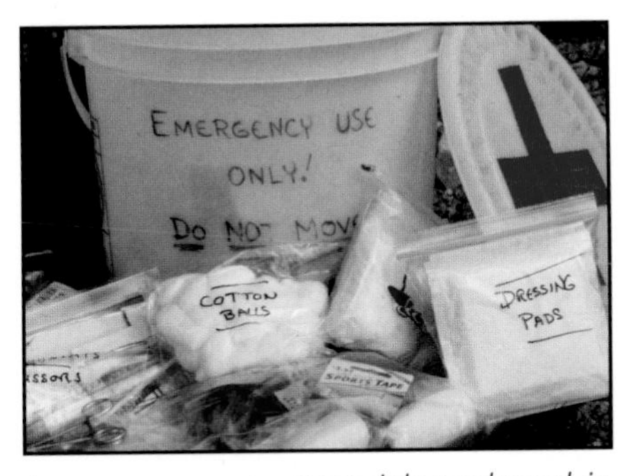

In an emergency, essentials packaged in ziplock bags are more easily located than items just tossed in haphazardly.

AND LLAMAS in large letters to differentiate it from our equine first aid kit (it's labeled HORSES). We keep the buckets in the house in a walk-in closet, and they're returned to their cubbyhole immediately after use—no exceptions.

One bucket contains emergency equipment such as lead ropes and halters ranging in size from our smallest small ruminant halter (it's an alpaca halter, but it fits our miniature sheep) up to horse-size halters. The horse halters are hand-tied rope versions to conserve space. When we must have an item in an emergency, we carry the bucket to the site and dump everything out so we can pick out what we need. The bucket also contains a fencing tool and a small length of wire for making impromptu fence repairs if an animal must be extracted from a fence.

The second bucket is organized using ziplock bags. Each bag is labeled in big black letters with an indelible marker.

One bag contains stuff for treating wounds: gauze sponges, Telfa pads, three rolls of VetWrap self-stick disposable bandage, a roll of two-and-a-half-inch sterile gauze bandage, one-inch-wide and two-inch-wide rolls of adhesive tape, a partial roll of duct tape, two heavy-duty sanitary napkins with wings (they can't be beat for applying pressure wraps to staunch bleeding), commercial blood stopper powder, a small bottle of Betadine Scrub, another of regular Betadine, and a twelve-ounce bottle of saline solution.

Another bag contains hardware: blunt-tipped bandage scissors, a hemostat (we like them better than tweezers), a flashlight, a stethoscope, and a digital thermometer.

A third contains basic medicines such as our wound treatments of choice (we like Schreiner's Herbal Solution); topical antibiotic eye ointment; and a full tube of Probios probiotic paste.

Packed alongside the rest of this stuff is a long package of roll cotton and two flat, wooden paint stirrers from the hardware store—combined with VetWraps, it's everything needed to splint a cria's broken leg.

We also store over-the-counter and prescription drugs for emergencies in plastic baskets in our pharmaceutical refrigerator so when we need an item we can grab the correct basket and run.

A separate, scaled-down first aid kit is stowed behind the seat of the truck.

Checking Vital Signs

Whether you contact your vet, your lama mentor, or post to a favorite e-mail list for advice, be ready to provide your sick or injured llama's or alpaca's vital signs

Advice from the Farm

Taking the Heat Off

The experts offer tips for preventing the number one killer of llamas and alpacas: heat stress.

The Boys Get Misty

"Here's a 'keeping the alpacas cool' tip I just discovered. Buy a 25-foot sprinkler/soaker hose at The Home Depot for about $20. Weave it through the fence about calf height from the ground, facing inside the pasture. Hook the hose to a water source, and turn it on: instant misting system for alpacas!

"The boys checked it out right away and even our guardian llama, Josh, stood downwind so he was getting the mist without looking like a 'sissy.' "

—Tina Cochran

Rolling Out the Wet Carpet

"Get remnants of indoor-outdoor carpet and wet it down. This cools the whole area as the fans blow air across it, and of course when the llamas kush on it they get that cooler air on their stomachs. They have 6-by-8-foot pieces at The Home Depot for $15. I originally held out for a bigger piece, but then I realized that smaller sections would be easier to manipulate, especially when I have to hose carpet off because somebody peed on it.

Lama - Pérou.

"Also, mark your water buckets. If the llamas aren't drinking from their regular buckets, mix up a smaller bucket of water with electrolytes in it and see if they will drink that. Some of mine like the orange flavor. The trouble is that you really need to change it every day, but if they'll drink it, it makes a huge difference in their hydration. I put the smaller bucket beside their normal water bucket so they have the option. Otherwise, those that don't like it won't drink at all if it's their only choice."

—Deb Logan

"After reading Deb's tip, I asked a local carpet installer for a big piece of used carpet from the Dumpster out behind his store. He was glad to get rid of it, and it works just fine!"

—Mary Collins

Temperature Taking

Take a lama's temperature in three easy steps:

1. Restrain the lama. It's best to restrain most llamas and alpacas in a chute, but if you don't have one, recruit a helper to steady the haltered lama by holding him against a wall or a fence.
2. Insert the business end of the lubricated thermometer (K-Y Jelly, Vaseline, and mineral oil are excellent lubes) about 2 inches into the lama's rectum. Avoid grasping the lama's tail to facilitate insertion; most llamas and alpacas don't like to have their tails handled, and doing so may trigger a rebellion.
3. Hold a glass thermometer in place for at least two minutes, or a digital model until it beeps.

(temperature, heart rate, and respiration) and to describe any symptoms in detail. The following are normal values for adult llamas and alpacas; crias' values run slightly higher. Keep in mind that normal values vary somewhat from individual to individual, so for best results, check and record each healthy animal's normal values to have on hand when you need them.

- Temperature: 99.5–102° F
- Heart rate: 60–90 beats per minute
- Respiration rate: 15–30 breaths per minute

Keep in mind also that external conditions may affect your readings. Lamas' temperatures rise slightly as the day progresses and may be a full degree higher on hot, sultry days. Extreme heat and fear or anger elevates pulse and respiration. Slightly elevated readings are sometimes the norm.

Temperature

The first thing your vet or mentor will ask is, "What is the animal's temperature?" To take it you'll need a rectal

Once you've taken your lama's temperature a few times, he'll get used to it and it will become a nonevent. This gorgeous alpaca doesn't even mind having his tail held!

thermometer. Veterinary models are best, but a digital rectal thermometer designed for humans works well, too.

Traditional veterinary thermometers are made of glass and have a ring on the end to which you can attach a string. Add an alligator clamp to one end of a fourteen-inch length of cord, and knot the other to the thermometer. That way you can attach the clamp to the hair of the patient's tail before inserting the thermometer and neither lose the thermometer inside the lama (yes, it can happen) nor drop and break it. Glass thermometers must be shaken down before and after every use; hold the thermometer firmly and shake it in a slinging motion to force the mercury back into the bulb.

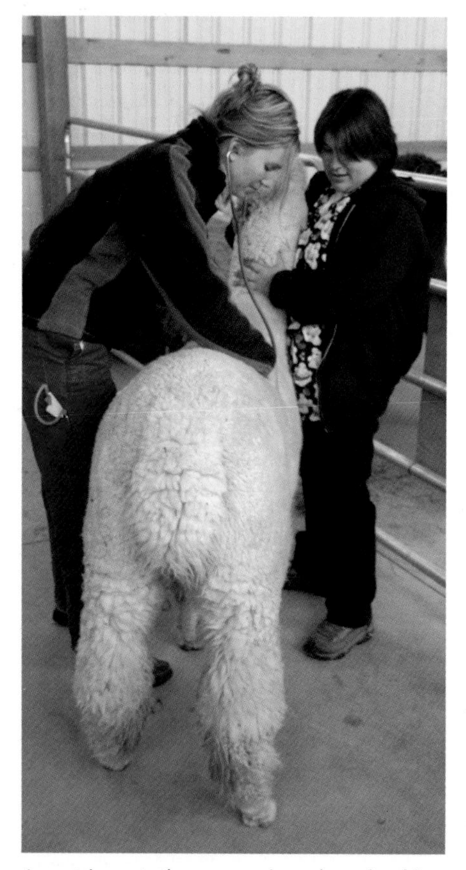

Invest in a stethoscope—it makes checking your lamas' heart rates quicker and easier!

Digital veterinary thermometers are better because they are are faster, they beep when finished, and they needn't be shaken down. Simply press a button and a digital model is reset. Digital thermometers are ideal when working with livestock.

After recording the reading, shake down the mercury (in glass models), clean the thermometer with an alcohol wipe, and return it to its case. Always store your thermometer at room temperature.

Heart Rate

The easiest way to check a lama's pulse is with a stethoscope. If you have one, simply listen and count the number of beats in fifteen seconds and multiply the total by four. If you don't have a stethoscope, take the lama's pulse by lightly pressing two fingers against the large artery on the inside of a rear leg up near the groin, counting the number of pulses in fifteen seconds, then multiply that number by four.

Respiration

Watch the rise and fall of your lama's rib cage (or if the animal isn't shorn, lightly place your fingers against its lower rib cage), and count the number of breaths the lama takes in fifteen seconds, then multiply by four.

Wound Care 101

Although a veterinarian should treat serious injuries of any kind, it's usually all right to handle minor wounds, cuts, and abrasions yourself. However, call the vet if:

- An injury is bleeding profusely.
- The injury is on or close to a tendon or a joint.
- A wound is contaminated by dirt or other debris.
- You've just discovered the wound, and it's already infected.
- It's a puncture wound (these are best treated by a vet).

Keep a plentiful supply of saline solution on hand for cleaning cuts and abrasions. If you're out, flush wounds using lots of cool, running water from a garden hose. Then apply a simple, mild disinfectant such as dilute Betadine solu-

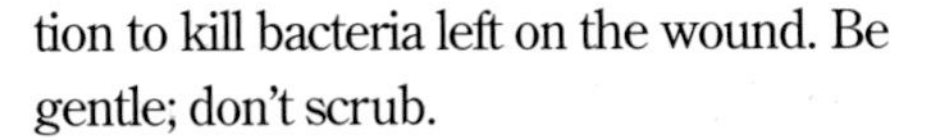

tion to kill bacteria left on the wound. Be gentle; don't scrub.

What to put on this nicely cleaned wound? In many cases, nothing. Clean, open, untreated wounds often heal better (and faster) than injuries coated with commercial preparations. If in doubt, ask your veterinarian for advice.

Just Shoot Me!

Every animal owner on a budget quickly becomes adept at giving shots. Here are the steps we use.

1. Select the correct pharmaceutical and reread the label. Don't omit this step.
2. Chose the smallest disposable syringe that will do the job; they're easier to handle than big, bulky syringes, especially for people with small hands. Disposable syringes are inexpensive and readily available, so use them but

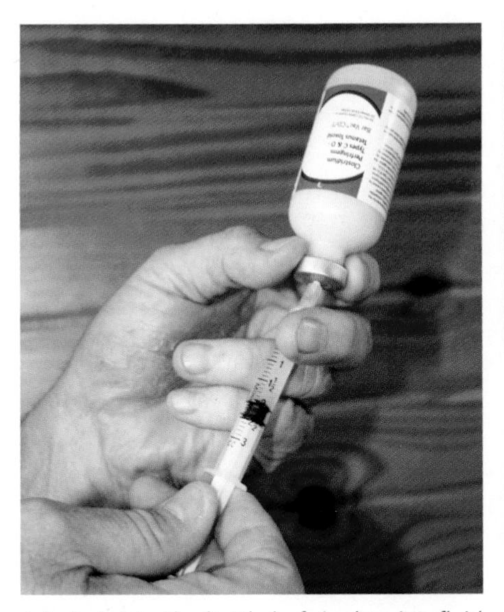

Inject air into the bottle before drawing fluid back into the syringe.

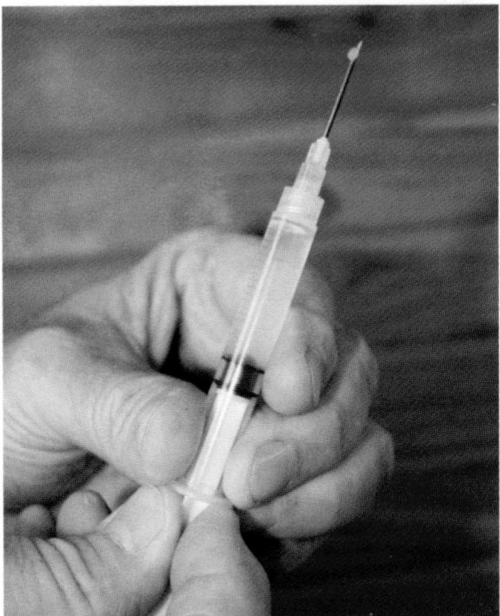

Then carefully eject any air bubbles before giving the injection.

never try to sterilize disposable syringes—boiling compromises their integrity.

3. Choose the correct needle for the job. Most antibiotics and vaccines should be injected subcutaneously (SQ, just under the skin) using a ½-inch, ¾-inch, or 1-inch needle. An 18- or 20-gauge needle is fine for adults; use 20- to 22-gauge needles for crias. Some antibiotics are very thick and when injected, the narrow carrier makes these injections sting; for these, some people prefer 16-gauge needles for adults so they can inject the fluid quickly, before their patient objects.
4. Select enough needles to do the job. You'll need a new needle for each animal, plus a transfer needle to stick through the rubber cap on each product. Using a new needle each time is much less painful for the animal, and doing so eliminates the possibility of transmitting disease via contaminated needles.
5. Restrain the llama or alpaca. Recruit enough helpers, or secure the animal in a chute.
6. Insert a new sterile transfer needle through the cap of each pharmaceutical bottle. As you draw each shot, attach the syringe to the transfer needle and suck the vaccine or drug into it, then detach the syringe and attach the needle you'll use to inject the pharmaceutical into your llama or alpaca. *Never* poke a used needle through the cap to draw vaccine or drugs.
7. If you're drawing 3 cc of fluid from the bottle, inject 4 cc of air (to avoid the considerable hassle of drawing fluid from a vacuum), then pull a tiny bit more than 3 cc of fluid into the syringe. Attach the needle you'll use to inject the pharmaceutical, then press out the excess fluid to remove any bubbles created as you drew out the vaccine or drug.

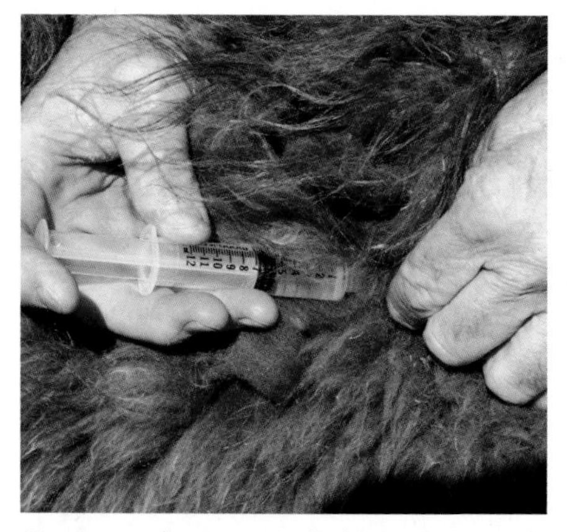

To give a subcutaneous (SQ) injection, simply pinch up a roll of skin and inject the vaccine or antibiotic into it.

It's good to follow a course of antibiotics with probiotics to help restore normal rumen function.

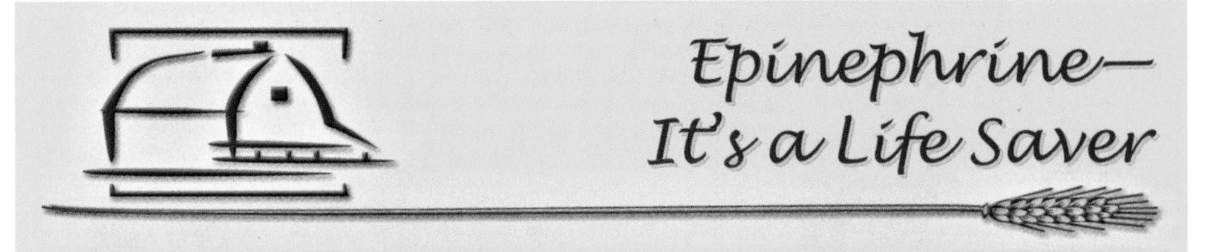

Epinephrine—It's a Life Saver

Epinephrine, also called adrenaline or epi, is a naturally occurring hormone and neurotransmitter manufactured by the adrenal glands. It was first isolated and identified in 1895 and artificially synthesized in 1904. Epinephrine is widely used to counteract the effects of anaphylactic shock, a serious and rapid allergic reaction that can kill.

Whenever you give an injection to an animal, no matter the product or amount injected, be ready to immediately administer epinephrine to counteract an unexpected anaphylactic reaction. If your llama or alpaca goes into anaphylaxis (symptoms include glassy eyes, increased salivation, sudden-onset labored breathing, disorientation, trembling, staggering, or collapse), you won't have time to race to the house to grab the epinephrine, and you might not even have time to fill a syringe. You must be ready to inject the epi *right then*.

We keep a dose of epinephrine drawn up in a syringe in the refrigerator. Sealed in an airtight container (we use a clean glass jar with a tight-fitting lid), it will keep as long as the expiration date on the epinephrine bottle. Take it with you every time you give a shot. Standard dosage is 1 cc per 100 pounds; don't overdose, as the drug causes the heart to race.

Until recently, injectable epinephrine was available over the counter, but now it's a prescription drug and available only through a vet. An alternative some animal owners use is over-the-counter Primatene Mist sprayed under the animal's tongue. Every fourteen or fifteen squirts of Primatene Mist contains the same amount of epinephrine as a 1 cc dose.

8. Select the best injection site. You want to inject the vaccine or drug where it will work well but won't injure your llama or alpaca. Preferred sites for SQ injections are into the loose skin of the animal's "armpit" and relatively loose skin in the area where the neck ends and the shoulder begins. Intramuscular injections are rarely used, but when you have to give one, the thick muscle of the "cheek" of your lama's rump is a winning location.

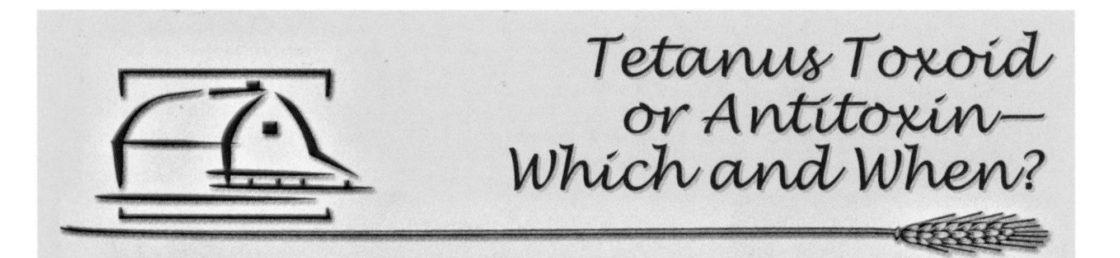

Tetanus Toxoid or Antitoxin—Which and When?

Some vaccines are available as toxoid and antitoxin versions, most notably vaccines against tetanus and enterotoxemia. The terms can be confusing, but this is what you need to know: toxoid imparts long-term immunity to llamas and alpacas. Toxoid is given in stages: an initial injection followed three weeks later by a second shot, then followed by boosters once or twice a year. Immunity, however, is not immediate, so if an unvaccinated animal is injured, *antitoxin* is injected to impart immediate but only temporary immunity.

Crias acquire immunity from disease via their vaccinated dams' colostrum, thus it's wise to boost pregnant females' immunity with toxoid vaccines four to six weeks prior to giving birth. The cria's own immune system begins to kick in around three months of age, so you need to start their series of tetanus toxoid injections at that time.

Tetanus antitoxin is administered (even to already vaccinated lamas) to males when they are castrated as well as to any llama or alpaca that suffers a deep puncture wound. The protection imparted by antitoxin is short lasting, a week or two at most. If a wound isn't healed at the end of seven to ten days, it will require revaccination with antitoxin.

9. When injecting a relatively large volume of fluid, break the dose into smaller increments and inject it into more than one injection site. (For example, don't inject more than ten cc of penicillin into a single injection site.) If you're unsure if this is necessary with the product you're using, ask your vet.
10. Carefully part the lama's wool and swab the area with alcohol (prepackaged alcohol swabs are easy to use). Never inject anything into damp skin or mud- or manure-encrusted skin.
11. To give an SQ injection, pinch up a tent of skin and slide the needle into it, parallel to the lama's body. Take care not to push the needle through the tented skin and out the opposite side or to prick the muscle mass below it. Slowly depress the plunger, withdraw the needle, and rub the injection site to help distribute the drug or vaccine.
12. To give an intramuscular injection, quickly but smoothly insert the needle deep into muscle mass, then aspirate (pull back on) the plunger one-quarter inch to see if you hit a vein. If blood gushes into the syringe, pull the needle out, taking great care not to inject any drug or vaccine as you do, and try another injection site.

How Much Is . . . ?

1 milliliter (1 ml) = 15 drops = 1 cubic centimeter (1 cc)
1 teaspoon (1 tsp) = 1 gram (1 gm) = 5 cubic centimeters (5 cc)
1 tablespoon (1 tbsp) = ½ ounce (½ oz) = 15 cubic centimeters (15 cc)
2 tablespoons (2 tbsp) = 1 ounce (1 oz) = 30 cubic centimeters (30 cc)
1 pint (1 pt) = 16 ounces (16 oz) = 480 cubic centimeters (480 cc)

Antibiotic Pros and Cons

Antibiotic overuse is a real and rapidly expanding problem. Many new livestock owners think, "I'm not going to use antibiotics on my animals." However, in some cases avoiding antibiotics simply isn't feasible. If your veterinarian says to use them, keep these tips in mind.

Follow directions to the letter. Inject the recommended dosages, and complete the series as directed, otherwise you'll kill only weaker pathogens; "supergerms" tend to survive.

Because antibiotics destroy good bacteria as well as bad, it's best to follow antibiotic treatment with oral probiotics such as Probios or Fastrack (the kinds labeled for ruminants) to restore your llama or alpaca's digestive system to good health.

Stop Disease in Its Tracks—Vaccinate!

Not all owners routinely vaccinate their llamas and alpacas, but they should. At minimum, vaccinate for tetanus. Depending on where you live and what sort of "bugs" haunt your locale, your vet will likely recommend additional vaccinations as well.

Using Health Care Products

To get the most out of the vaccines, medicines, and dewormers you pay for, choose and use them to your best advantage. When you buy a new product, read the label and store it accordingly. Some drugs need to be refrigerated, others kept out of direct light; some should be shaken before being administered and are ineffective if that step isn't taken. Note any warnings: some pharmaceuticals cause abortion if used to treat pregnant females.

Reread the label prior to using any product; it's easy to forget particulars between uses. If a product was stored incorrectly, discard it. Check the expiration date before administering a drug or

A drenching gun makes drenching (giving liquid medicine to) a lama less stressful.

a vaccine, and throw it out if it's outdated. Don't buy the large, economy size if you can't use it before it goes bad.

Liquid medicines and some dewormers are given orally as drenches. Liquids can be administered using nozzle-nosed, catheter-tip syringes (not the kind you use to give shots), but the most efficient way to drench is with a commercial dose syringe. Whichever you use, wear protective glasses or goggles and old clothing; many lamas strongly object to oral drenching, paste dewormers, and the like. If sufficiently aggravated, these lamas spit.

To drench a llama or alpaca, elevate its head using one hand under its chin—not too high, but let gravity help you a little bit. Insert the nozzle of the syringe between the back teeth and cheek (this way the lama is less likely to aspirate part of the drench) and *slowly* depress the plunger, giving the animal ample time to swallow.

When giving paste-type horse wormers or gelled medications, instead of squirting the substance between the cheek and teeth, deposit it as far back on the lama's tongue as you possibly can. In either case, keep the nose slightly elevated until the animal visibly swallows.

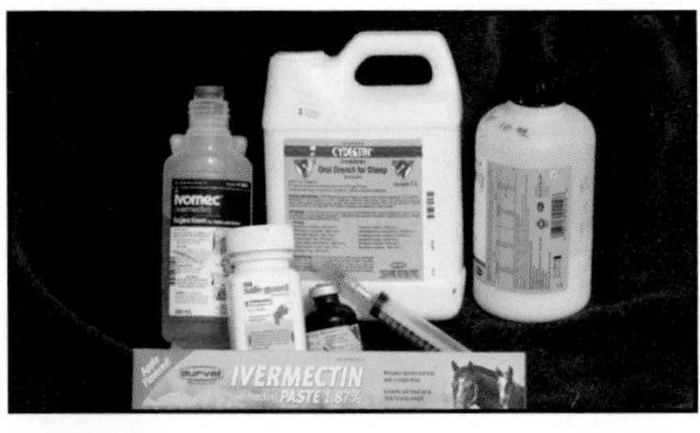

There are no deworming products specifically labeled for lamas. Ask your vet which ones he recommends you use.

The Worms Go In, the Worms Go Out

Llamas and alpacas harbor an array of internal parasites (known to most of us simply as "worms"), but only a few are (usually) serious enough to damage their hosts' bodies. Even so, any type of heavy worm infestation increases feed costs and causes stress.

Llamas and alpacas are prone to most of the internal and external parasites that afflict sheep, goats, and cattle. You'll never vanquish these parasites once and for all, especially if your lamas live with other host species, but you can and must control them. The best way to do this is to work with your veterinarian to discover which parasites your llamas, alpacas, and similar host species have and which dewormers are still effective on your farm.

Dewormer-Resistant Worms

Dewormer-resistant internal parasites present a growing problem not only throughout North America but also all around the world. For decades, veterinarians told sheep, goat, and cattle producers to deworm all of their stock at the same time and to rotate dewormers to reduce drug resistance. Since few owners weighed their livestock before deworming, many also underdosed their animals. In this manner, farm animals were exposed to all of the available anthel-

Barber pole worm larvae may be lurking in that perfect sea of grass, so a proper deworming program is an essential part of lama keeping.

mintics, often in doses too low to effectively kill worms. Because of this, the weaker worms died, but the strong ones survived. Now dewormer-resistant "superworms" have evolved to the point that most anthelmintics have lost their punch—and the drug companies aren't developing new products to replace them.

Two major species that plague sheep, goats, and cattle and that are commonly found in llamas and alpacas too are the barber pole worm and the brown stomach worm.

Barber Pole Worms

The barber pole worm (*Haemonchus contortus*, also called the twisted wireworm) is a member of the gastrointestinal trichostrongyles family, a group of nematodes sometimes called roundworms. The barber pole worm's life cycle is typical of most nematode species.

A single female barber pole worm lays between 5,000 and 10,000 eggs *a day*. While living in the stomach of cattle, lamas, sheep, or goats, these females release eggs that become encapsulated in the animal's droppings, and then the eggs fall to the ground. In one or two days, the eggs hatch and, still encased in the animal's droppings, begin their lives as first-stage larvae. They soon molt into second-stage larvae. Both stages feed on bacteria eliminated in the droppings, while storing energy they need to survive later on.

They molt again, becoming third-stage larvae. Then, providing the droppings stay moist, they emerge and begin "questing" for a host. If the droppings are too hard, the third-stage larvae become arrested, meaning they stay put for one to three months until moisture softens the droppings and they can survive in the outside world.

As third-stage larvae still encased in their second-stage skins, they feed on stored energy. Each larva climbs up a blade of grass and waits to be accidentally eaten by some grazing beast. Exactly how long the larvae can survive depends on air temperature and the amount of energy they've stored, but thirty to ninety days is the norm.

When a lama, sheep, goat, or cow happens along and consumes the blade of grass the third-stage larvae are clinging to, the larvae set up house in the animal's stomach, where they molt, becoming fourth-stage larvae. Then if conditions are favorable (based on a number of variables, including air temperature, greening of grass, rain following a drought, and estrogen spurts of a female

Crias are especially susceptible to coccidiosis.

host giving birth), fourth-stage larvae molt into adults. If conditions are unfavorable, they remain arrested as fourth-stage larvae and wait for conditions to improve.

Two to three weeks after entering their host as third-stage larvae, adults breed, females lay eggs, and the three-week cycle begins again.

Barber pole worms feed on blood. When a host harbors lots of these worms, it faces life-threatening anemia. A mere 1,000 barber pole worms can rob their host of up to a pint and a half of blood per day; an infestation of roughly 10,000 barber pole worms can kill a sheep or a goat. Even if they don't kill their host, heavy barber pole worm infestations do irreparable internal damage to their hosts' stomachs, resulting in poor feed conversion for the rest of the animals' lives.

Brown Stomach Worm

Brown stomach worm (*Teladorsagia circumcincta*, formerly called *Ostertagia circumcincta*), another member of the gastrointestinal trichostrongyles family, also lives in its host's stomach, where it feeds on nutrients harvested from the stomach's mucous lining. This permanently damages the organ, which in turn affects its ability to digest nutrients, so infested animals fail to thrive.

Additional Gastrointestinal Nematodes

Although infestations of the other gastrointestinal nematodes aren't generally as serious as infestations by "the big two," they still adversely affect their hosts' health and well-being. Infestations involving more than one worm species (especially in conjunction with the barber pole worm or the brown stomach worm) are especially likely to result in permanent damage. Several other internal parasites that infect llamas and alpacas include *coccidia*, tapeworms, liver flukes, and meningeal worms.

Coccidia (*Eimeria* sp.). Coccidiosis is a potentially fatal, highly contagious disease affecting llamas and alpacas, particularly crias. Other animal species are afflicted by coccidiosis, but Eimeria protozoa are species specific, meaning your llamas and alpacas can't catch chicken or dog coccidiosis (and vice versa). However, a single infected lama can shed thousands of microscopic coccidial oocysts (the spores by which coccidia reproduce) in its droppings every day. If another llama or alpaca accidentally ingests a sporulated (mature) oocyst, it can become ill a week or two later. As oocysts multiply and parasitize the gut, they destroy their new host's intestinal lining. Without immediate aggressive treatment some young animals, particularly unweaned crias, die; others develop chronic coccidiosis, culminating in stunted growth.

Symptoms include watery diarrhea (scouring), sometimes containing blood or mucus; listlessness; poor appetite; and abdominal pain.

Regular dewormers don't kill coccidia. Sulfa drugs are usually the treatment of choice, and electrolytes such as RE-SORB, Pedialyte, or Gatorade can be given orally to rehydrate stricken crias. Consult your veterinarian or lama mentor for particulars; this is a life-threatening situation, so don't delay.

Tapeworm (*Moniezia* sp.). If you see living tapeworm segments (resembling grains of white rice) in recently deposited llama or alpaca droppings, don't be unduly alarmed. A major infestation may slow growth in crias, but unless an intestinal blockage occurs, tapeworms aren't a primary problem.

Liver fluke (*Fasciola hepatica*). Liver flukes pose a serious problem in the wetlands of the Southeast and sometimes farther north as well. Liver flukes can live in the host up to ten years. They feed off the bile duct lining and over time can cause enough irritation to cause scarring and cirrhosis of the liver, possibly leading to death. Symptoms of liver fluke infestation include anemia, severe weight loss, and low milk yields. Liver fluke requires an intermediate host (a snail) to reach infection stage.

Meningeal worm (*Parelaphostrongylus tenuis*). Meningeal worm is sometimes called deer worm or meningeal deer worm because its natural host is the white-tailed deer. Llamas and alpacas (as well as sheep, goats, and wild species such as moose) are at risk wherever whitetail deer are present. Ground-dwelling slugs and snails are the intermediate host between deer and other ruminant species. Although the worm doesn't bother its natural host to any great degree, in other species larvae migrate throughout to the host's spinal cord and brain causing rear leg weakness, staggering gait, hypermetria (exaggerated stepping motions), circling, gradual weight loss, and paralysis leading to death. It's a major problem for llama and alpaca owners in white-tailed deer country.

In endemic areas, most vets prescribe off-label, preventative ivermectin injections at thirty-day intervals year-round. Treatment is difficult and generally unsuccessful, so if you live in proximity to white-tailed deer, discuss meningeal worm with your vet; this is important!

The Solution to Drug Resistance Is . . . ?

Unfortunately, there is no easy fix. With no new dewormers on the horizon, we must somehow prevent further anthelmintics resistance but also rid our llamas and alpacas of worms. There are several steps you can take.

Avoid across-the-board deworming. It's estimated that in most herds, 20 to 30 percent of lamas, goats, and sheep carry the majority of the worms and shed the majority of the eggs.

Nematodes are now resistant to many dewormers, so it's best to have fecal tests run to make certain the ones you use are still effective on your farm.

These animals require frequent deworming; the others don't.

Don't rotate dewormers every time you dose your llamas and alpacas. If an anthelmintic is working, use it for at least a year or until it loses its effectiveness.

Every few weeks, gently pull a lower eyelid away from each animal's face and examine its mucous membrane. It should be red to dark pink. Pale pink to white membranes indicate that the animal is anemic and needs to be dewormed right away.

Check animals' membranes carefully about one week after it rains following a period of drought. Arrested larvae bloom under these conditions.

Deworm all females immediately after they give birth, when changing estrogen levels cause arrested larvae to proliferate.

Weigh your lama and dose accordingly. Never underdose!

With few exceptions, dewormers should be delivered by mouth. No matter what the labels say, never use them as pour-ons. Use a dose syringe to deposit oral dewormers on the back of the llama or alpaca's tongue, and then elevate the chin until the animal swallows.

Make sure your deworming program is working. Collect droppings from animals you suspect are wormy, place each donor's offering in its own labeled plastic bag, and take your samples to the vet. (Only fresh-from-the-donor droppings should be used, preferably pellets that haven't touched the ground.) The vet will prepare a solution from each sample, examine it under a microscope, and count the number of worm eggs on each slide. Ten days after deworming each donor, collect more samples and run new tests. There should be a 90 to 95 percent reduction in egg number per gram of feces. If not, drug resistance is brewing on your farm.

Check with your vet or lama mentor before using products labeled for other species, and *always* use the correct dosage per product per llama or alpaca.

Don't introduce drug-resistant worms to your farm. Quarantine and deworm all incoming llamas, alpacas, and similarly susceptible livestock (cattle, sheep, and goats), using two products from different drug classes according to your vet's or lama mentor's recommendations. Give new acquisitions time to shed superworm eggs by keeping them indoors or in a small pen and away from any pasture for a minimum of forty-eight hours.

Feed the animals well. Well-fed, otherwise healthy llamas and alpacas are more parasite-resistant than ailing lamas are.

Discuss parasite control strategies with your vet or county extension agent; set up a first class worm-control program tailor made for your llamas' and alpacas' specific needs.

External Creepy-Crawlies

External parasites such as flies, ticks, and lice feed on body tissue, including blood, skin, and hair, and they can transmit diseases from sick to healthy animals.

Nose Bots

The nose bot is a hairy, yellowish fly about the size of a common housefly; nose botflies are often mistaken for bees. Female nose botflies lay their first-stage larvae in the nostrils of lamas, sheep, and goats. A single female botfly may deposit as many as 500 larvae in the nostrils of unwary lamas, sheep, and goats. They migrate up the nasal passages and feed on mucus and mucous membranes. Hosts lose their appetites and shake their heads; there is often an opaque discharge from infested nostrils. Currently, ivermectin is the only effective treatment for nose bots.

Lice

Lice spend their entire lives on a single host. Both immature and adult stages suck blood and feed on skin. Louse-infested animals rub against all available surfaces, resulting in raw patches of skin and loss of hair. Weight loss occurs if infested animals don't eat. In severe cases, the loss of blood to sucking lice can lead to serious anemia.

There are two types of lice: sucking lice (these pierce the host's skin to suck their blood) and biting lice (these have chewing mouthparts and feed on particles of hair and scabs). Lice are generally spread via direct contact, often when infested animals join an existing herd. Populations vary seasonally. Most sucking and biting lice proliferate during autumn and reach peak numbers in late winter or early spring. Summer infestations are rare.

Louse control is difficult because pesticides kill lice but not their eggs. Since eggs of most species hatch eight to twelve days after pesticide application, retreatment is necessary two or three weeks following the first application.

Mites

Itch or mange mites feed on skin surface or burrow into it, making teensy, winding tunnels from one-tenth of an inch to one-inch long. Fluid discharged at the mouth of each tunnel dries and forms scabs. Mites also secrete a toxin that causes intense itching. Infested animals rub and scratch themselves raw. Infestations are highly contagious; if one llama has mites, treat the whole herd.

Itch mites are highly contagious; if you find some, treat your whole herd!

Open Wide and Say Cheeeeese

Depending on the ages, sexes, and jaw structures of your llamas and alpacas, you may or may not have to trim their teeth. If you do, here's what you need to know.

This gorgeous, older llama has overgrown incisors.

Fighting Teeth

Intact male llamas begin growing six sharp fighting teeth, or fangs, two on the top gum and one on the bottom on each side of the mouth, at roughly eighteen to twenty-four months of age. A few females and even some geldings get them as well but not to the degree that intact males do. Unless these are blunted or removed, males use them on one another, often inflicting considerable damage on their opponents.

Extracting fighting teeth is risky and often results in injuries to the jaw. Instead, fighting teeth are usually sawed off at gum level at intervals until they stop growing at full maturity.

Your veterinarian can saw these off at the gum line using a Dremel-type tool or flexible cutting wire called OB wire, designed to slice hard surfaces without damaging soft tissue. You can also do it yourself, but have a vet or experienced person show you how the first time.

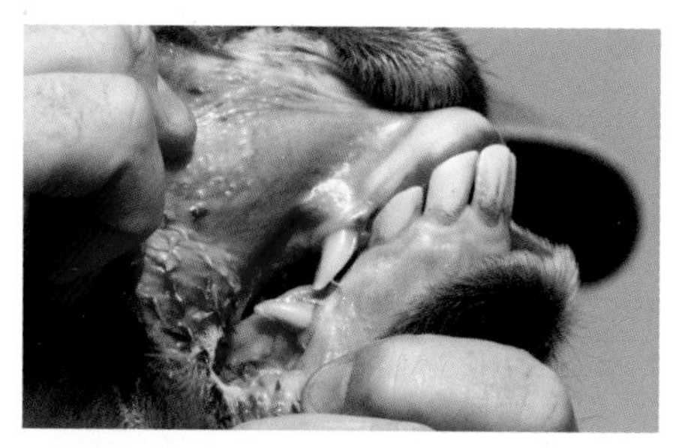

Lamas don't have upper teeth; instead, their lower teeth meet their upper dental pads. Notice Bandit's large, sharp fighting teeth. Adult males (even geldings) have fighting teeth unless they're sawed off or surgically removed.

Overgrown Incisors

Aged llamas and alpacas as well as younger lamas with poor occlusion sometimes need to have protruding front teeth trimmed. Use a battery-powered Dremel tool with a diamond wheel to trim back overgrown incisors, keeping in mind that you can remove a lot of tooth without injuring the lama. However, never grind all the way down to root level because exposed roots are prone to dental infections. Whether using OB wire to trim fighting teeth or a Dremel to grind back incisors, stop frequently to let the tooth you're working on cool down. A cardinal rule: never cut llama or alpaca teeth with side cutters or nippers because they will shatter! Use the right tools, and take your time to do the job correctly.

Trimming Tootsies

Llama and alpacas need their toenails trimmed from time to time, but the frequency varies greatly from lama to lama. Some need the chore performed every six weeks; a few kept on rocky soil don't need it at all.

Unless you teach them otherwise, lamas dearly hate to pick up their feet. Lamas are prey animals; they instinctively know that if they can't run when danger beckons, they're dead meat. In addition, male llamas, even geldings, fight (and play fight) by biting at one another's lower legs. When you try to pick up that same leg, they resist. Is it any wonder why?

Thus, most llamas and alpacas require understanding (and usually a handling chute) at nail-trimming time. Even so, many resist. And it doesn't hurt to wear spit-protective clothing the first time you trim a new lama's nails.

To pick up a lama's foot, start with your hand on its shoulder or haunch and run it down the leg until you reach the foot. Give an oral signal ("foot," "leg," or "up" work equally well), grasp the foot, and then pick it up. If you practice before nail-trimming day and consistently offer your lama a tasty food reward for picking up his foot (give the treat while his foot is in the air, not after you've placed it back on the ground), so much the better.

Trimming lamas' toenails is an ongoing part of lama maintenance.

You need the proper tool to trim nails correctly, and it's the sort of spring handled hoof cutter used to trim the hooves of sheep and goats. Most farm stores carry them; if yours doesn't, buy them from one of the suppliers listed at the back of this book.

To manicure a lama's nails, carefully trim along the sides, avoiding soft tissue. If you've never done it before, be conservative; if you "quick" your lama (cut into soft tissue, making the lama bleed), the animal will be a lot less cooperative next time. When the sides are neatly cut back, snip back the tip by trimming straight across, perpendicular to the rest of the nail, and then shape it by carefully rounding the sharp edges.

If you do quick your llama or alpaca, treat the wound with styptic powder (cornstarch will do in a pinch) or Betadine solution. Don't cut as far next time!

The Rest of the Story

If you live in a hot, humid climate, you'll probably need to shear your lamas every year.

CHAPTER SEVEN

Breeding Llamas

Sooner or later, many llama owners opt to breed a cria or two, and some breed as a monetary venture. Before you join these ranks, however, here are some things to consider. Since female alpacas are extremely costly at the time this book goes to press, we're assuming few hobby farmers keep them. However, if you do, these principles apply to breeding alpacas, too.

Why do you want to breed your female? If you think it's an inexpensive way to add more llamas to your herd, think again. There are countless fine llamas looking for homes, so rescuing or buying one is far more cost-effective than paying a stud fee and board at the male owner's farm, feeding your female through gestation and lactation, and raising a cria to adulthood. (And you can train and enjoy the adult you purchase right away.) Everyone loves a cute, fuzzy cria, but what happens when the charmer grows into a rowdy teenager just lounging around, eating, and taking up space? Will you sell him? Are you willing to teach him ground manners, geld him, and screen buyers so he finds a good home?

Is your female of breeding quality, and if she is, are you willing to pay what it costs to breed her to a comparable male? Be brutally honest—if your female isn't of breeding quality, don't breed her! If she is, realize it's costly to breed her to a quality male. Prebreeding vet work, stud fees, boarding fees, and postbreeding vet work add up quickly. If your female doesn't conceive, at worst you're out your money, and at best you'll have to rebreed until she gets pregnant.

Do you think you'll make big bucks breeding llamas? A few breeders do, although most of them would question the term *big bucks*. In other words, it's difficult to break even breeding llamas, so if you do breed them, it had better be

Did You Know?

- All South American camelids have seventy-four chromosomes and are cross-fertile, producing fertile first-generation offspring.
- The most common cross is a llama-to-alpaca mating, producing offspring intermediate in size, body characteristics, and fiber quality. None of the other crosses are common, although the structural similarities between llamas and guanacos make identification of hybrids difficult, so there could be more than we think.

for love. Keep in mind that no matter what you breed you must begin with top-of-the-line breeding stock, maintain it under optimal conditions, plan breedings with utmost care, and aggressively market your product. Startup and ongoing expenses will be substantial. We'll talk more about this in chapter nine.

I'm not trying to discourage you; I'm simply asking you to give the topic serious thought before breeding your female. Then, if you do the homework and you still want to breed your own cria, this is what you need to know before you do.

Choosing a Male

The first step in breeding is finding the perfect male for your female. If she has flaws—and all animals do—the male should be strong where she is weak. Never breed two individuals that share the same flaws.

Where to Look

Search for stud males using the same resources cited in chapter two; people who have crias for sale may own just the male you'd like to breed to. Once you've located two or three potential boyfriends for your female, contact their owners and ask the following questions.

What sorts of fees are involved? Some male owners include a certain number of days of board for your female in the stud fee. If so, how many and what will happen after that? If you'll be paying across the board for your female's keep, how much will it cost and what does the fee include?

Where and how will your female be kept? Will she have her own shelter and exercise area or be turned out in a strange herd to fend for herself? What happens if she's injured while she's away being bred?

Stud males should have heavy bone and correct conformation.

Does the male owner require specific vaccinations, tests, or health papers before your female sets foot on his farm? What about specific vaccinations that your female might not already have?

If you like what you hear, ask the stud male owners to mail you copies of their breeding contracts and any additional information you'd like to know about their males. When they come, read the contracts carefully and compare your options, then arrange to see the males you are still considering.

When you visit, ask to see where your female will be kept. Are you comfortable leaving her there? If not, is the male owner willing to provide other accommodations, perhaps at a greater price?

Look the male over carefully and examine his get (offspring), especially the ones whose dams resemble your female. Both conformationwise and dispositionwise, are they the sort of crias you want to produce?

After examining all contenders, make your choice and get a signed breeding contract outlining all the specifics of this breeding. That's step one!

A Stud Male of Your Own

Unless you breed a number of females every year, you probably won't need your own intact male llama. Quality males are expensive. If you buy one, he'll require separate housing and substantial fencing; you'll feed him year-round, even if you breed only one female a year. By using outside males, you can choose the ideal mate for each of your females.

If you want to own a stud male, keep in mind that in most states male lamas, stallions, jacks, bulls, rams, buck goats, and the like are considered dangerous animals. You are liable for any damage your male inflicts on any human or animal, even if your land is posted "no trespassing."

And if you stand your male to outside females and hope to show a profit, plan to market his services aggressively. Investigate the legal angles and hire a lawyer to draw up a breeding contract tailored specifically for your needs. Discuss stud male handling and the merits of pasture breeding versus hand breeding with experienced breeders who stand stud males, and then build your facilities accordingly.

Only the best males should be used for breeding, like this striking stud male at Klein Himmel Llamas.

Did You Know?

- Rama, the world's first cama, was born on January 14, 1998, at the Dubai Camel Reproduction Centre in the United Arab Emirates. Rama's size is midway between that of his mother, Smokey, a guanaco, and his father, Musehan, a dromedary racing camel. Rama lacks a camel's hump and has the toes of a lama, as well as the long tail and short ears of a camel.
- In 2002 a second cama was born, this time a female named Kamilah, meaning "perfect." Another female (Jamila) followed in 2004 and another male (Rocky) in 2005.
- The camas were conceived via artificial insemination, with ovulation induced by hormone injection. More than fifty attempts to impregnate lama ovum with camel sperm were tried, resulting in fifteen conceptions, six of which led to live births. The other two camas died as infants.
- Unlike true hybrids, lamas and camels share the same number of chromosomes (seventy-four), so the camas may well reproduce.

The Birds and the Bees, Lama Style

Most female mammals "come in heat," meaning they periodically become receptive to males of their species, and at that time and that time only, conception is likely to occur. Not so the female lama. Instead, she is sexually receptive most of the time. Lamas are "induced ovulators," meaning they ovulate in response to being bred; they share this unusual breeding physiology with cats, ferrets, rabbits, mink, skunks, short-tailed shrews, and thirteen-lined ground squirrels.

Lamas mate lying down with the female in a kush position and the male facing the same direction. He orgles (it's a distinctive vocalization that sounds a lot like loud, noisy gargling) for the duration of their mating session, which can last from fifteen minutes to an hour or more. Lamas are dribble ejaculators, thus it takes a considerable amount of time to do the deed.

A typical mating session begins with the orgling, amorous male, his penis extruded for mating, trying his best to convince his paramour to lie down. If she's willing, she does; if she's already pregnant, she'll "spit him off," meaning she'll actively repulse his advances.

Let's assume she's interested. She kushes and he mounts, gradually scooting his body closer to hers. When he's close enough, he clutches her sides with his forelegs and snakes his long, thin penis through her vagina and all the way into her uterus, where he dribbles sperm until the mating ends.

Sound (orgling) and stimulation (penetration and the tight grasp of the male's front legs) causes the female's brain to release hormones that, in turn, cause her to ovulate approximately thirty to forty hours after mating. If all goes well, she conceives.

Seven days after mating, owners test for possible pregnancy by reintroducing their male to his ladylove. If she spits him off, this is good news. Further testing and spitoffs should occur at weekly intervals up to day twenty-eight after breeding. At this point she can be considered

Already orgling, Bandit encourages Sis to kush.

Sis seems quite bemused while Bandit gets down to business.

pregnant—or, to be absolutely certain, tested further.

Ultrasound is the best way to ascertain that a female is pregnant. Lama-savvy vets prefer to ultrasound between forty-five and sixty days postmating, as by this time a viable pregnancy can be easily seen. Or your vet may prefer running a blood test to check the female's progesterone levels. Both tests work quite well.

Did You Know?

While Americans simply say a female lama gives birth, in some countries the act of giving birth is called unpacking or criation. Likewise, while we just call them females and males, elsewhere the South American terms *macho* (for intact males) and *hembra* (for females) are used.

Before You Breed Your Llamas

Make sure your female is old enough to breed. Maidens (young females) are generally bred at fourteen to eighteen months of age, but if your female isn't nearing physical maturity, it's always better to wait until she's older.

Consider the time of year your female will give birth. Llama gestation is said to last 350 to 355 days, but the truth of the matter is, normal crias can be born anywhere between 315 to 375 days after conception. You don't want your cria to be born in deep winter cold nor sultry summer heat, because crias' bodies, like those of most newborns, aren't terribly adept at thermoregulation. In addition, third-trimester mama llamas are far more prone to heat stress than are their

Females should be in good health before breeding.

peers, so in hot, humid climates, breeding for summertime births is very unwise.

Be sure your female is healthy and in good flesh, neither too fat nor too thin, when she leaves for her date with the stud male. Then, when she comes home, keep the following tips in mind.

Minimize traveling during early pregnancy, especially if traveling stresses your female. Get her home and keep her there; if she needs to see a vet, pay the extra dollars for a farm call.

Don't let your female get fat. Almost 85 percent of fetal growth occurs during the last trimester of pregnancy, so she doesn't require supplemental feeding until late in the game. Feed good grass hay during early gestation, making certain she has a plentiful supply of clean water and a properly balanced mineral available at all times. Unless she's grossly overweight, begin feeding a judicious amount of grain about six weeks before her first due date.

Continue worming her on her usual schedule, using safe wormers based on ivermectin or fenbendazole (or any other wormer your vet approves of). If Valbazen is part of your worming strategy, however, don't use it to worm your pregnant female; Valbazen has been strongly implicated in spontaneous abortions and birth defects, a lesson I learned firsthand.

Four to six weeks prior to her first estimated delivery date, she should receive at minimum a CD/T booster (for enterotoxemia and tetanus protection). That way she's sure to pass antibodies against these diseases to her cria via her colostrum (first milk). Since a cria's own immune system doesn't start functioning until it's three or four months old, it

A Question of Breeding

The experts talk about breeding and raising lamas.

Consider the Reasons

"As with any animal, people need to really consider reasons for breeding. These animals live a very long time! We're already seeing an influx of 'fiber male' alpacas into rescue. This is not surprising, especially in challenging economic times, as too many males quickly become 'overhead' in a business predicated primarily on breeding."

—*Deb Logan*

Registration Is Important

"There are some breeders who want you to think that males and geldings are worthless. The alpaca industry sometimes loses sight of the fiber aspect of alpacas. The impression that there are more alpacas than needed to supply demand comes from the erroneous idea that only top-quality breeding stock has any value. That's simply not so.

"One thing: registration is extremely important if you plan to sell alpacas. And even if you aren't planning on breeding, buying registered stock is one way to ensure there will be a market for your alpacas if you decide they aren't for you or if your situation changes."

—*Tina Cochran*

Homes for the Little Ones

"The farm where we bought Cash and Cary makes sure that even their low-end crias find good homes by feeding and worming them well and training them to lead and load in a trailer before they're sold. Cash and Cary are only soon-to-be-gelded fiber alpacas, but they were raised the same way as the farm's expensive female crias. I like that!"

—*Mary Collins*

needs that passive transfer of antibodies from his dam to survive.

At the same time that we boost our late gestation female livestock, we inject them with a selenium and vitamin E supplement called BoSe. This is a common (and essential) practice in selenium-deficient parts of the country; check with your veterinarian or county extension agent to see if it's needed in your locale.

And be sure to keep her toenails trimmed; she's carrying an extra load, you know.

A healthy new cria is a joyous addition to your farm and worth the time and preparation involved.

If you own your own male, remove him from the herd during your female's last trimester. As birth approaches, the cria's placenta will begin to produce more estrogen, which sometimes makes the female smell as though she's receptive. If a male mates her (submissive females often kush whenever a male demands it), she could lose her pregnancy. And by all means remove him before your female gives birth! Males have been known to try to breed females in the act of giving birth or soon after, and if that happens, your cria could be killed.

Preparing for Delivery

At least a month before your female's first due date, assemble a birthing kit, making certain you have everything you need. Set up one or more birthing areas (in the barn or in easily accessible paddocks or pens), depending on the number of females expecting crias; don't assume they'll take turns because they often don't.

If you put your female in peril by breeding her, *be there* when she gives birth. Female llamas die every year when the presence of a human attendant could have made the difference between life and death, if only to call a veterinarian. Once that baby is ready to be delivered, it's your responsibility to be there if something goes wrong.

A Well-Stocked Birthing Kit

We pack our birthing supplies in a Rubbermaid toolbox stool. It's roomy, it has a lift-out tray for small items, it's easy to tote to the barn, and it's much nicer to sit on than an overturned five-gallon bucket. Others swear by tackle boxes, toolboxes, and even backpacks. Whichever you choose, your kit should contain:

- A bottle of 2 to 7 percent iodine (used to dip the newborn's umbilical cord).
- A shot glass (to hold cord-dipping iodine).
- A digital veterinary thermometer.
- Several shoulder-length, preferably sterile, obstetric gloves.
- Plenty of obstetric lube. We like a brand called Super Lube, formulated for assisting ewes at lambing time.
- Betadine Surgical Scrub (for cleaning females' nether parts prior to assisting).
- Dental floss (to tie off an umbilical cord that drips blood).
- A bulb syringe (the kind designed for human infants).
- Sharp scissors (you never know when you might need them; sterilize them and keep them in a sealed ziplock bag).
- A sharp pocketknife (ditto).
- A hemostat (ditto).
- A halter and lead (you won't want to leave a female in distress to scare one up if you need it).
- Several large, soft, terrycloth towels (and if it's cold outside, add a hand-held hairdryer to the kit).
- A clock or a watch.

We keep our birthing kit in a Rubbermaid step stool so we always have comfortable seating for the big event.

Stow all the small items in sealed ziplock bags so you can find them quickly in the heat of an emergency. Label them in big letters using a permanent marker.

You will also need a cell phone (or a groundline to the barn), one or two clean buckets, a ready supply of warm water, and access to a van or trailer in case you have to rush your female to the vet.

Are You Ready for the Birth?

Although most females give birth without assistance, unless your vet is certain to be no more than fifteen minutes away from your farm at birthing time, be ready to help out if the need arises. It sounds scary and you may think you can't do it, but you can—otherwise, your female, her cria, or both could die. If the worst should happen and you have to intervene, here's what you need to know.

Keep your fingernails clipped short and filed (you won't have time for a manicure when an emergency arises).

Know how to determine what configuration a malpositioned cria is in and precisely how to correct it. If you think you might forget, photocopy and laminate instructions to keep in your birthing kit.

Practice. Teach your fingers to "see." Borrow a pile of plush toy animals. Place one in a fairly close-fitting paper bag. Without looking, stick your hand in the bag and figure out what you're feeling. Switch animals. You'd be surprised how helpful this exercise can be.

Program appropriate numbers into your cell phone so you can call at least two good vets at the touch of a button.

Stall, Pen, or Pasture?

Most breeders who monitor their females at birthing time prefer that they give birth in containment, such as in a roomy stall in the barn or a grassy pen.

Wherever your female gives birth (especially if you won't be there when she does it), make certain she delivers her cria under hygienic conditions. The area should be roomy enough for the female to walk around and roll in comfortably. There must be no protrusions along the walls (nails, splinters, bucket hooks) for a newborn cria to get hurt on. The area should provide privacy for the female, but she should be able to see other llamas in the distance so she doesn't stress out. And it should offer an unobtrusive spot from which an attendant can observe the birth. Watering receptacles should be small enough or hung high enough so your female can't deliver her cria into them (since nearly all females give birth while standing up or squatting, it happens) and the newborn cria, as it learns to handle its wobbly legs, won't tumble headfirst into a water source and drown.

Are We There Yet?

As delivery day approaches (and be aware that llamas sometimes give birth weeks before their official due date), start watching for some or all of these signs.

Two or three weeks before the female's cria is due, her udder may begin to enlarge, although some females don't "make bag" until a day or so before giving birth. She may roll more than usual, and she'll likely spend more time kushed, with or without the company of her friends. You'll notice she looks increasingly larger as her underbelly starts to expand.

A few days before her cria arrives, you'll notice her returning time and

Make certain your female has a safe, familiar place in which to give birth.

again to one certain spot—she's scouting out a place to give birth. Her udder is probably bigger, her vulva looks more relaxed and definitely swollen, and you might even see a bit of discharge at her vulva as her cervical plug begins to come loose.

Two to eight hours before giving birth, she'll probably leave the herd, possibly in the company of a friend. She'll become very restless. She's obviously in great discomfort. She'll hum or even moan while lying down, getting up, and then lying down again, over and over. She'll intermittently lift her tail up, and she'll make many trips to the dung pile where she'll strain but produce few, if any, "beans."

When birth is imminent, she'll hum more stridently and spend more time rolling, getting up, and lying down again. Her back will stay arched, her tail will be up continuously, and she may turn to look under her tail. Her tail may be very wet, signaling that her waters have broken. When that happens, here comes baby!

Female lamas generally give birth standing up; a few prefer to squat or give birth lying down. All three positions are perfectly normal.

The first thing to appear at the female's vulva is a fluid-filled, water balloonlike sac called the chorion, one of two separate sacs that enclose the developing fetus within its mother's womb (the other is the amnion). Either or both sacs can burst within the female or externally as the cria is born.

In a normal front-feet-first, diving position delivery, first one foot appears inside the chorion (or directly in the vulva if the chorion has already burst), followed by another foot a few inches behind the first one, and then the cria's nose. Once the head is delivered, the rest is generally easy. However, the front half of the cria may dangle from its standing or squatting mother's birth canal for fifteen to twenty minutes before dropping to the ground. This is normal, so don't panic when it happens.

Did You Know?

Nearly all lama births occur between 7 a.m. and 2 p.m. This is nature's way of assuring South American crias are up and dry before frigid nights in the Andes begin. While nighttime birthing isn't necessarily cause for alarm, more dystocias (difficult labors) do occur at night. If your female gives birth late in the afternoon or during the night, call your vet and ask him to please stand by.

When a newborn cria seems unusually weak (it can't hold its neck up or stand at one or two hours of age), it has a weak or missing suck reflex (put your finger in its mouth to find out), it has very short fleece or no teeth (the tips of its central incisors should be breaking through the gums), or its ears are quite floppy, it may be premature. Call your vet! And to be prepared in case it happens to your cria, visit the Alpaca Association New Zealand Web site (see Resources) and download their free Neonatals Paddock Card (to find it, click on *Alpaca Information*, then scroll down the page); the material is outstanding!

When You Must Help

If something goes awry, call your vet right away. If he can't arrive in a reasonable amount of time, he may want to talk you through handling the dystocia yourself. If you can't get a hold of your vet, try another. If no one can arrive in time, take a deep breath, stay calm, and handle the situation the best you can. It's scary, but with care and determination, most dystocias are correctable.

Because of anatomical differences, you can't pull a cria the way you'd pull a calf. If you must pull, use *lots* of lube and pull only during contractions. Grasp the legs, preferably above the pasterns but below the knees, then pull. Don't pull straight back, pull out and down in a gentle curve toward the female's hocks.

Before entering the female, make absolutely certain your fingernails are short and that you've removed your watch and rings. Wash her vulva using warm water and mild soap or a product such as Betadine Surgical Scrub. Pull on an OB glove if you have one; if you don't, scrub your hand and forearm with whatever you used to clean the female's vulva. Then, liberally slather the glove or your hand and arm with lube. Now pinch your fingers together and gently work your hand into the vulva.

The female will be doing her best to push the cria out, even if it's hopelessly stuck. When contractions hit, hold still; when they let up, work quickly but carefully until the next one occurs.

Determine which parts of the cria are present in the birth canal. Closing your eyes and moving your awareness to your fingertips will help. If the cria's toes point upward and the big joint above it bends away from the direction the toes are pointing, it's a foreleg. If the toes point down and the major joint bends in the same direction, it's a hind leg. Follow each leg to the shoulder or groin if you can, making sure the parts you're feeling belong to the same cria. (It's very, very unusual, but lamas occasionally have twins.) When you're certain they do, try to manipulate the cria into a normal birthing position, then help pull the cria out. By this time the female will probably be too exhausted to do it by herself.

This female's herd mates gather around her as first stage labor begins.

Most lamas give birth in a standing or squatting position.

When the cria's hips clear the birth canal, it will topple to the ground.

Lamas are very protective mothers and may become aggressive in the presence of their crias.

The inside of a female llama is extremely fragile, and if you or the cria tears her, she'll almost certainly die. When repositioning the cria, work carefully and deliberately; the female llama's life depends on your gentle technique.

In a breech presentation, two feet followed by hocks appear. Because the umbilical cord is pressed against the rim of the dam's pelvis during this delivery, it's wise to *gently* pull the cria once its hips appear.

In a full breech presentation, the cria comes butt first with its hind legs tucked forward. It's definitely best not to try to reposition the cria yourself. However, if you must reposition this cria, try to push it forward, work your hand past its body (it's a tight squeeze), and grasp one hock. Raise the hock up, and rotate it out away from the body. While holding the leg in that position, use the little and ring fingers of the same hand to try to work the foot back and into normal position; repeat on the opposite side. The umbilical cord will be pinched, so pull the cria as quickly as you safely can.

If the cria presents with one foreleg back, push back just far enough to allow you to cup your hand around the trailing front foot and gently pull it forward.

Occasionally, two front legs but no head appears. Because of the length of the cria's neck, this is an especially difficult position to correct. To give yourself some room, gently push the cria back into the vagina as far as you can, and only then work the head up or around into position.

In dystocias, hip locks are fairly common. What it means is that one edge of the cria's pelvis is hooked on the edge of its dam's pelvis and the cria is stuck. If a cria dangles from its dam's birth canal for longer than twenty minutes, it's best to help the delivery proceed. Gently pull the cria down and to the right, then down to the left, then to the right again, and so on until you've "walked" the cria's pelvis through the female's pelvic structure.

Elbow locks are fairly common, too. In normal deliveries, one leg is delivered so that the tip of one front foot is approximately at the level of the fetlock of the other front leg. If the leading foreleg is too far ahead of the trailing foreleg, the

Possible Cria Birth Positions

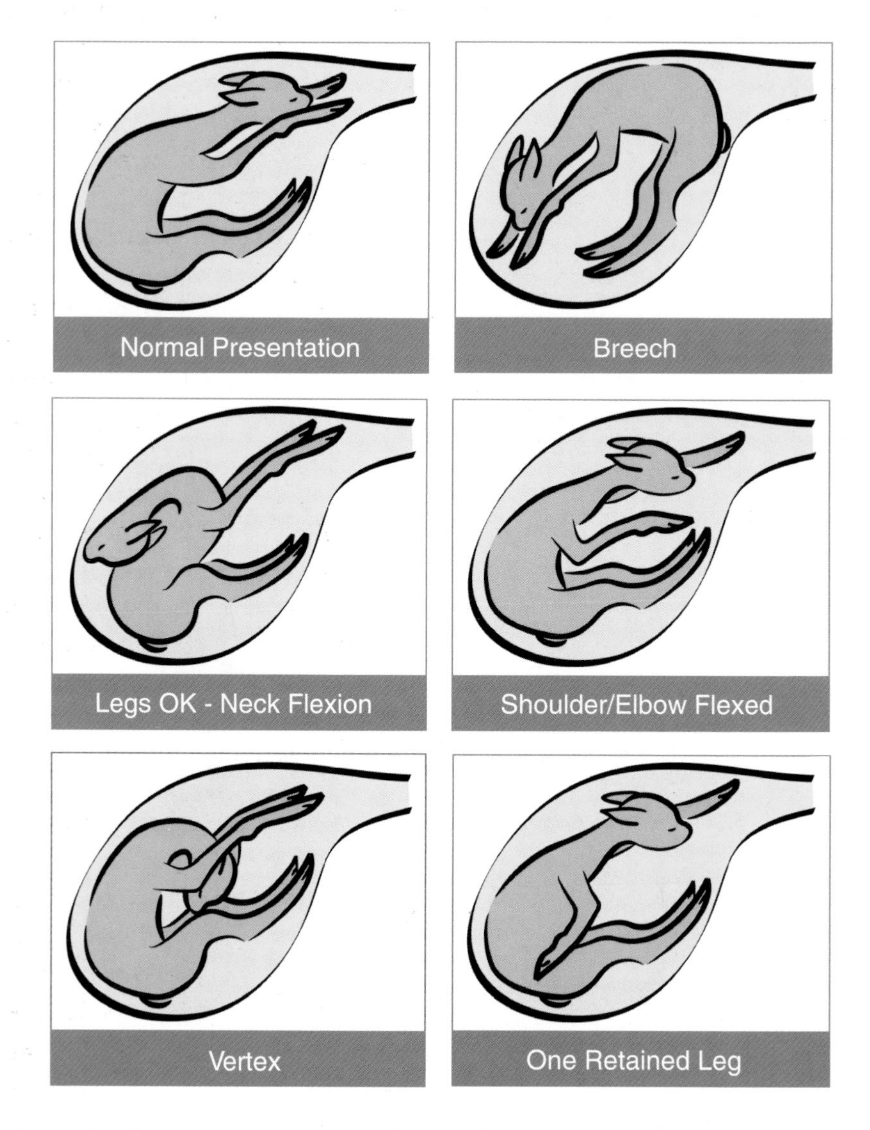

trailing leg begins to flex, and the elbow becomes lodged against the female's pelvis. Elbow lock is easily corrected by pushing the cria back, grasping the foot of the trailing leg, and then pulling it forward into the proper position so the delivery can proceed.

Postbirthing Procedures

Once the cria arrives, immediately remove birthing fluids from its nose by stripping your fingers along the sides of its face. If the cria is struggling to breathe, use the bulb syringe from your birthing kit to suck fluids out of

Did You Know?

A cria's umbilical cord usually ruptures a few inches from the animal's body. If it's more than three inches long, use your sterilized scissors to trim it to an inch and a half in length before dipping it in iodine for the first time. If the cord continues to drip blood, clamp your sterile hemostat on it an inch below the cria's belly only until the bleeding stops, or use clean dental floss to tie it off prior to dipping. If it drips clear fluid, call your vet—clear fluid could be urine, and if it is, your cria requires veterinary attention.

Every cria is born with thick, rubbery coverings on its feet. These are designed to prevent its sharp toenails from injuring its mom during birth. You don't have to remove them; baby will quickly wear them off all by itself.

the nostrils, then tickle the inside of the nose with a piece of straw or hay. If that doesn't work, you can try to give it CPR. Realistically, if a newborn's heart isn't beating, you probably won't be able to save it. If the heart is beating, this procedure might just do the trick.

Check the cria's mouth and clear it of any foreign material, then lay it flat on its right side. Close the mouth and pinch the lower nostril shut. Now place your mouth over the upper nostril and blow a lungful of air into the lungs until you see the chest rise. Remove your mouth from the cria's nostril and let the air escape. Repeat for about three minutes, giving the cria one breath roughly every three seconds. If it doesn't work in this amount of time, quit. The cria would likely be brain damaged if you manage to resuscitate it beyond this point.

Once you're positive the cria is breathing well, remove any stringy birthing tissue clinging to its body and use the towels or hairdryer from your birthing kit to dry it.

When it's dry, weigh it (full-term llama crias generally weigh at least twenty pounds). You can do this with an accurate bathroom scale by weighing yourself first, then holding the cria and subtracting the difference. Write it down; you'll want to reweigh the cria every day for the first ten days or so to make certain that it's eating enough and gaining weight. A healthy cria may lose a few ounces for the first day or two after birth but should start gaining about eight ounces per day after day two.

Next, "dip, strip, and sip": Dip the cord in iodine. Fill a shot glass or empty film canister with iodine, hold the container to the cria's belly so the cord is completely submersed, then tip the animal back to effect full coverage; don't omit this important step (and be sure to repeat it twice more before the cria is eight hours old). Then, strip the female's teats to make certain they aren't plugged with natural body wax and that she indeed has milk. Finally, settle back, relax, and observe the little family until the cria has sipped its first meal of colostrum.

Colostrum

Colostrum is the thick, yellowish milk produced by female lamas beginning about two weeks prior to the birth and ending a few hours after her cria arrives. It's loaded with important nutrients, but more important, colostrum contains vital immunoglobulins (antibodies) that crias need to survive. They must ingest an adequate amount of colostrum within six to eight hours of birth because their gastrointestinal systems can absorb the immunoglobulins in colostrum only during a short window of time—no longer than twelve to twenty-four hours after birth.

Immunoglobulins are proteins used by the immune system to identify and neutralize foreign bodies such as bacteria and viruses. Crias that don't ingest colostrum lack immunity to diseases such as tetanus and enterotoxemia until their own immune systems kick in. Crias also need the nourishment in colostrum to prevent hypoglycemia (low blood sugar). Crias have little or no fat stores and require frequent meals. Going several hours without colostrum or milk can leave a newborn cria very weak and unable to stand. Unless crias ingest colostrum or they're given a suitable substitute, such as a plasma transfer, they rarely survive; it's that important.

If a cria can't nurse from its dam, but she's still available, milk her and bottle-feed her colostrum to the cria in small, frequent increments until he's ingested 10 percent of his total body weight in fluid. Never feed milk or milk replacers during a cria's first twelve to twenty-four hours of life.

If you can't harvest colostrum from the cria's dam, fresh or frozen colostrum from another female lama on your farm will do the trick. Barring that, goat, sheep, or even cow colostrum is acceptable. If no colostrum is available, the only alternative is intravenous treatment with immunoglobulin-rich plasma. Call your vet!

Since every cria needs colostrum to survive (barring the expensive plasma transfusion), it pays to keep some on hand in your freezer. Properly frozen colostrum (lama, goat, sheep, or cow) stays good for up to one year. Quick-freeze it in 4-ounce increments in double-layered ziplock sandwich bags. Avoid storing it in self-defrosting freezers; constant thaw and refreeze cycles affect its integrity. *Never* microwave colostrum, as this kills the protective antibodies;

Alpaca crias are cute beyond words! The non-slip mats in this birthing stall help this little guy stand up straight and tall.

Did You Know?

- Llama milk has a higher sugar count (6.5 percent) and lower fat content (2.7 percent) than milk from other ruminants. It has more calcium and less sodium, potassium, and chloride, but the concentration of trace minerals is similar to that of cow milk.
- Female camelids have four teats. Their udders are somewhat similar to those of deer.
- Supernumerary (extra) teats are more common in llamas and alpacas than in the Old World camels.
- A "milk mustache" is a sure indication that a newborn cria is nursing.

instead, immerse the container in hot water until the colostrum registers 100 to 102 degrees Fahrenheit (check it with a thermometer to be certain).

Make certain your cria is actually nursing. If the tail is up, the cria probably attached. If not, get under its mama to see what the cria is doing and if necessary, help it latch on. Then, get out of the stall or birthing pen, and allow the little family to bond.

Don't Forget the Placenta

Within an hour or so after your cria arrives, your female will go into third-phase labor and deliver the placenta (or afterbirth), the tissues in which her cria developed inside her uterus. It is vitally important that these tissues come out promptly and intact. If she doesn't pass the placenta within eight hours, call your vet without delay.

Encourage your female to work at expelling the placenta by getting out of her stall and reducing distractions to a bare minimum, but *don't* pull on exposed portions of the placenta in an effort to help! This can cause her uterus to prolapse (pull out and dangle outside of her body); a prolapsed uterus is a major emergency.

Always wear protective gloves when handling the placenta. It's unusual, but infected placental tissues can transmit diseases such as brucellosis (undulant fever) to humans who handle them.

After it's expelled, spread the placenta out and check it over; it should be shaped like a pair of panty hose with a main section and two legs ending in round, closed feet. If you think something is missing, place it in a bucket of cool water, cover it, and call the vet. If everything seems all right, dispose of the placenta by burning or burying it—don't let your dogs or cats eat it!

Two adorable youngsters at Klein Himmel Llamas stay toasty warm in matching, homemade cria coats.

Can you imagine a prettier picture than this Klein Himmel Llamas appaloosa mama with her colorfully clad cria at her side?

Taking Care of Baby

Once your cria is safely here, you'll want to do everything you can to help it grow up happy, healthy, and strong. Here are some things you should know.

Meconium Happens

The first manure a cria passes is a black, tarry substance called meconium. (Once it's ejected, the cria will make soft, yellow-tan stools.) Occasionally, crias have a good deal of trouble passing meconium. If your cria seems to be straining, give it a child-size Fleet enema. Lubricate the tip, and please be *gentle*.

Keeping Baby Warm and Dry

Neonatal crias are delicate creatures; you mustn't allow one to get wet or chilled. Keep your cria inside during inclement weather, and if it inadvertently gets soaked, bring it in, dry it off, and warm it up as quickly as you can.

Wintertime neonates usually require supplementary heat to stay warm. However, resist the urge to hang a heat lamp in your barn! Improperly hung heat lamps burn down barns, and it's not an unusual occurrence. For a much better and safer alternative, fit your cria with a body covering to hold in heat. Commercially made cria coats are nice, but dog sweaters and homemade cria coverings work well, too. To make an easy, effective cria cozy for pennies, buy a used children's wool cardigan sweater at your favorite used-a-bit store, snip off the sleeves, and fit it to your cria so it buttons along his spine. In addition, deeply bed the maternity stall with straw so the cria can hunker down into it to stay warm.

Neonatal crias are all neck and legs!

Getting Along with Mama

Some usually easygoing females become fiercely protective once their crias are born. Don't assume you can sashay back in the stall an hour or two after baby was born and mama has gotten back her wind.

No matter how relaxed your female usually is, don't get between her and her newborn cria. Before you reenter the stall, plan an escape route and try to have an assistant standing by, just in case.

Don't make eye contact with any female who has a newborn cria—to lamas, eye contact is a sign of aggression, and a brand-new mom might take offense.

If she does and she only spits, be a good sport. Spit happens. She's protecting her newborn babe the best way she knows how.

Whether you handle your growing cria is your call. Some people do; others say it contributes to the development of aberrant behavior syndrome, so they leave their crias alone until weaning time. If you do it, go back and reread chapter three to make certain you aren't creating a brat. In our opinion, a moderate amount of training, stressing manners as part of the curriculum, is indeed a very good thing.

Healthy, well-grown crias can be weaned at four to six months of age, although to allow for maximum mental and emotional development, six months or even more is a better plan.

When weaning day arrives, remove either mama or cria from the main herd, and keep the pair separate for at least one full month. Each should be provided with at least one companion, whether another llama or a friendly sheep, goat, or another laid-back barnyard friend.

CHAPTER EIGHT

More Great Lama Activities

What can you do with llamas and alpacas? Let me count the ways! Llamas and alpacas are ideal hobby farm pets, it's true, but there are lots of other great things to do with lamas. Here are a few to consider.

Adopt a Lama

Pet lamas, pack llamas, guardian llamas, even fiber and agility llamas or alpacas—why not adopt some? There are lots of good reasons to do it, not the least of which is that hundreds of worthy llamas (and an increasing number of alpacas) are looking for forever homes just like yours.

Llama overpopulation is a real and scary problem. During the past few decades, indiscriminate breeders and "dabblers"—people who keep a male and a few females just to see cute crias born on their farms—flooded the market with a glut of inexpensive, unregistered llamas; this in turn decreased the demand for responsible breeders' excess stock. The result: there are now more llamas than responsible homes that want them. Now excess male alpacas are heading the same direction.

What Rescue Is and Isn't

Many prospective lama owners assume animals in rescue are second-rate survivors of neglect or abuse situations, but nothing could be further from the truth. According to statistics provided by Southeast Llama Rescue, one of the largest and most active llama rescues in the world, 80 percent of their intake is composed of healthy, sometimes show quality, and often registered llamas whose owners simply

Although you won't find breeding alpacas through rescues, adoption is a fine way to add geldings to your herd.

couldn't keep them any longer. The organization's motto is "Things change."

People surrender lamas due to catastrophic life changes such as death in the family, divorce, loss of income, and military deployment. They don't want the lamas they love to go to dubious homes or through sale barns where they might end up as meat. So they give them to a rescue to place. That's where you come in.

And the other 20 percent? Some are, indeed, survivors of neglect. These receive the feed, care, and medical attention they need to be restored to full health before they're offered for adoption to new homes. Others, especially intact males, are sometimes surrendered for behavior issues. These males are castrated, then kept in knowledgeable foster homes and worked with until they're deemed fit for adoption.

When reputable rescues take in llamas and alpacas, the animals are wormed, their toenails trimmed, any dental issues addressed, and their coats sheared, if required. The rescues observe each lama closely over a period of time and carefully evaluate each one's behavior patterns, quirks, and eccentricities. If an individual is to be offered as a guardian llama, it is exposed to a variety of livestock (usually sheep, goats, and alpacas) before being considered for placement.

The bottom line: by adopting from a reputable rescue you'll eliminate most unknowns. And, if an adopted lama doesn't work out, the rescue gladly takes it back.

Rescue, however, is not the place to obtain breeding lamas. Every responsible rescue requires castration of males (gelding is done prior to adoption when age isn't an issue), and females are placed under no-breeding contracts.

Paying an adoption fee does not mean you own the animal you adopt. If at any time you choose not to keep llamas or alpacas adopted through a rescue, they must be surrendered back to the organization from which they came or to an individual or agency approved by the original rescue. You must also agree to allow rescue personnel or their representatives to visit the lamas periodically to ascertain that they are, indeed, being properly cared for.

Adoption is the perfect choice if:

- You are unsure of your ability to choose a healthy, well-behaved lama on your own. (Rescue lamas are carefully vetted prior to placement.)

- You want assurance that your lamas will be provided for in the event of your own changes of circumstance. (Rescue lamas automatically reenter the system when you can no longer provide for them.)
- You're seeking a llama suited for a specific job. (Herd guardians are pretested; fiber llamas are made available to fiber homes; pack llamas and even llamas with show win histories are sometimes available.)
- You're a good-hearted person who wants to give a home to lamas that need one.

Adopting Llamas

Although adoption policies vary from organization to organization, Southeast Llama Rescue protocols (SELR) are typical. In fact, if you live in one of the areas the group serves (eighteen states at this writing), contact it for details; sixty-three SELR llamas are looking for new homes as I pen these words.

This striking, ultra-banana-eared Klein Himmel Llamas female and her pretty cria both have white-tipped ears.

To adopt llamas from SELR you must have, at minimum, one-half acre of safely fenced exercise space for every two llamas you adopt and sufficient shelter to house them. You must agree to provide proper feed, everyday care, and veterinary attention as needed. After your application is processed, an SELR representative will inspect your facilities, help you select llamas specific for your needs, and help you find a mentor in your area if you don't already have one.

SELR places llamas in pairs unless the adopter already has lamas or the llama will be serving as a livestock guardian (and has already shown aptitude as such). The adoption fee is currently $250 per llama, which includes a properly fitting halter and lead, a book on llama care, and transportation for the llamas to your facility.

Foster Care

SELR and most other reputable lama rescues need foster homes for lamas awaiting adoption. Foster homes must comply with the above criteria and agree to provide everyday care and feed at their own expense. SELR pays for emergency medical expenses and gives foster caregivers first chance to adopt the llamas in their care.

Other Ways You Can Help

If you can't adopt or foster a needy llama or alpaca, you can help in other ways. Ask your favorite lama charity for their wish list. Donations of feed, medical supplies, dewormers, halters, leads, and other barn supplies—and money—are always needed, as are volunteers to transport or work with the lamas. Donations to bona fide rescues are tax deductible.

Go Llama Packing

When I was young, I loved to go backpacking. Now, like others of the baby boomer generation, the thought of lugging a cumbersome, fully loaded pack up hill and down dale isn't nearly as appealing. However, old folks, folks with bad backs, young folks with children in tow, folks like you and me, can still enjoy long, rewarding jaunts into the backcountry with the help of sturdy pack animals to tote their gear.

Why Pack with Llamas?

Llamas make the best pack animals bar none; they were bred for this job. Unlike the hooves of burros and pack goats, the llamas' large, wide feet and soft, rounded toe pads mold to the trail surface. The

Lamas have been an integral part of Aymara life for millennia—and they still are.

llama's unique toe pads provide traction in mud and snow and stability when scaling rocky surfaces, and they have minimum impact on fragile, backcountry terrain.

Llamas' droppings are environmentally friendly. No llama deposits big, damp, fly-attractive droppings along the trail in the manner of horses, burros, and mules. Instead, they wait until they find—or you make them—a potty pile to do the deed.

They can pack a lot of gear—or even a lightweight child should the need arise. A healthy, well-conditioned, mature llama (four years old or more) of either sex can easily carry 35 percent of its body weight in gear for short distances and 25 percent of its weight over longer hauls. A fit, 400-pound llama can carry 100 pounds.

Furthermore, it's easy to get llamas to the trailhead; no cumbersome horse trailers are needed. Load your gear and a pair of llamas in your full-size van or SUV, and off you go—it's that easy!

You needn't break the bank to outfit your llama for packing. A top-flight soft pack and panniers runs $300 to $400; a quality, sawbuck-style saddle and panniers, fully complete, $400 to $600.

Finally, if you already have a good-natured, sound llama accustomed to being handled, that walks quietly on a lead, you can go recreational packing with the llama you own. Llamas adapt quickly to packing, and there several great books and videos available (see the Resources section) to teach you and your llama the basics. Or buy a seasoned pack llama and head for more challenging trails; many professional llama packers sell them.

Competition Pack Llamas

In addition to camping with your pack llamas, you can show them in packing classes at llama shows or compete in pack llama trials.

Alpaca Llama Show Association (ALSA) lama shows (see the Resources section) offer a fine array of packing classes for llamas and alpacas. In these events, each lama, wearing a pack system or training pack and a set amount of weight depending on its age, is led through a course of eight to ten obstacles that simulate conditions encountered on the trail.

ALSA also sanctions pack trials that are often held in conjunction with approved shows. In these, llamas two years of age and older compete over field courses to earn titles, such as:

- Recognition of Merit Basic Packer—completing two basic field courses each at a distance of 4 miles in a maximum of three hours, carrying 25 pounds exclusive of saddle and rigging;
- Recognition of Merit Advanced Packer—for completing three advanced field courses each at a distance of 8 miles in a maximum of six hours, carrying 40 pounds;
- Recognition of Merit Master Packer—on the conclusion of completing at least four master field courses on three different trails, each at a distance of 12 miles in a maximum of eight hours, carrying 65 pounds.

Another organization, the Pack Llama Trial Association (see the Resources section), confers the titles Basic Pack Llama (BPL), Advanced Pack Llama (APL), and Master Pack Llama (MPL) to llamas that complete the required number of courses at each level.

Drive Your Lamas

What could be more fun than lazing along a country road in a cart behind a handsome llama or a team of fine alpacas? Many people do. You can, too!

A llama or a team of alpacas fitted with an appropriate harness and hitched to a lama-friendly cart can pull an adult or two for hours, especially when the driver includes longish periods of walking between each segment of trotting or cantering.

Readers who drive equines and goats are probably thinking, cantering? Yes indeed. Lamas tire quickly at a trot, but due to body conformation, they canter easily and readily in harness. The shafts (the pair of poles that run parallel to a single driving animal's sides and appear to connect the animal to the cart) or poles (the long pole running between two animals in a team) of lama carts are longer than those of horse, pony, and goat carts for precisely this reason.

We're not going to describe training a llama or a pair of alpacas to drive because driving is not something best learned from books. Improperly harnessed or hitched animals driven by inexperienced drivers are (serious) accidents waiting to happen. Take lessons from a driving instructor, attend a lama-driving clinic, or buy and absorb a video devoted to lama driving. I can't say this too emphatically: you *must* know the ins and outs of driving before you train your own lamas to harness! Once you do, here are so some things to consider.

It's always best to buy equipment designed for lamas rather than trying to make horse or goat equipment fit. Lamas are built differently than equines are. Consider the way horses and lamas hold their heads. Equines break at the poll (the neck joint right behind the ears) and tuck back their chins; lamas hold their heads up high with their heads almost parallel to the ground. For this reason, bits and lamas don't mix. Instead, you'll need a special lama driving halter. A goat harness generally utilizes a driving halter and can be fairly easily adapted for a small llama or single alpaca. Keep in mind, however, that a lama under roughly forty inches tall at the withers isn't big enough to pull significant weight by itself; smaller lamas do better driven as teams.

Choose a well-balanced, easy-entry cart when driving green (inexperienced) lamas; you'll want to be able to exit easily to go to your lama's aid if something goes wrong. Remember those longer shafts: they're important. In addition, insist on a cart with a single tree (the piv-

Did You Know?

- There are three movie star llamas: Kuzco, the emperor llama—from Disney's feature-length cartoon *The Emperor's New Groove*; Tina, the title character's pet llama in *Napoleon Dynamite*; and the Pushme-Pullyu of *Dr. Dolittle* fame.
- Scores of versions of "The Llama Song" by Burton Earny ("Here's a llama, there's a llama, and another little llama, fuzzy llama, funny llama, llama llama, duck") are among the most often-viewed videos at YouTube.

oted horizontal crossbar on the cart, directly behind the lama, to which the harness traces are to be attached); cheap carts sometimes don't have one. Without a single tree to allow for movement, the harness will chafe your animal's shoulders.

Choose your driving llamas or alpacas with care. Prospective driving lamas should stand quietly when tied, allow themselves to be touched all over, and pick up their feet with relative ease (in case your lama cuts or bruises a foot pad while out driving). The animal must be alert but basically easygoing, like and trust people, and be willing to try new activities. Females and males (gelded or intact) work equally well in harness. Although ground training can begin at eighteen months of age, driving lamas should be mentally mature and at least two years of age before being hitched to a cart.

You can drive just for fun but also consider participating in Alpaca Llama Show Association driving classes for llamas (at this writing, alpacas aren't allowed to show in driving classes). There are two sections for llamas: open pleasure driving (divided into single-hitch and team-hitch classes) and obstacle driving.

In pleasure driving classes, entries demonstrate their proficiency by moving at three distinct speeds; stopping, standing quietly, and backing on command; and individually driving a figure eight pattern if the judge so chooses.

Obstacle driving classes demonstrate a llama's training, the driver's skill, and the coordination of both llama and driver. Each entry enters the ring individually and navigates eight obstacles. These could include maneuvering around large objects such as gates and farm objects; navigating water obstacles; backing through poles; executing changes of gaits on command; passing animate objects such as dogs and children playing; dropping a piece of mail in a mailbox; or standing still while the driver dismounts, walks to the llama's head, and then remounts and drives away.

Showing Llamas and Alpacas

No matter what type of llama or alpaca you own, there are ALSA show classes written just for you. Or, if you're a 4-H club member, you can

Showing lamas is fun—and great publicity if you breed or if you stand a male at stud.

participate in 4-H lama classes and show your lamas at 4-H venues.

For a peek into the world of showing llamas and alpacas, visit the Alpaca Llama Show Association Web site (see Resources), read the various pages, and then download a free copy of the current ALSA rulebook. Both llamas and alpacas are shown in a vast array of ALSA halter classes (these are judged on the animal's conformation, soundness, and fiber), showmanship classes (in which the handler's ability to show his lama is judged), performance classes (from packing to driving to public relations events; even agility classes for alpacas), and fleece classes judged either on or off of the lama. Besides mainline classes, many shows choose classes from ALSA's optional list, too. These include fun events such as costume (in which both lama and handler wear costumes with a common theme), dam and cria (mamas shown with fuzzy babies by their sides), ground driving (for driving llamas that haven't advanced to pulling a cart), and drive and pack (in which the same llama is first judged pulling a cart and then unharnessed and shown as a pack llama).

In 4-H, children between the ages of nine and nineteen participate in age-appropriate activities ranging from keeping accurate records on a llama they own or lease, to attending training seminars sponsored by their state or county 4-H lama group, to exhibiting their project lamas or their lamas' fiber at the county fair.

Get a Llama to Guard Your Goats, Sheep, or Alpacas

Predation of sheep, goats, and alpacas by dogs, coyotes, and larger predators such as mountain lions and bears is a growing problem. Many people who keep small hoofed stock or alpacas also keep donkeys or livestock guardian dogs to protect their animals. In many instances, a guard llama can neatly do the trick and do it well.

In 1990, researchers at Iowa State University polled 145 sheep producers in five western states to determine the effectiveness of llamas for reducing dog and coyote predation. The producers reported an average annual loss of 21 percent of their ewes and lambs before adding a llama to the mix and only 7 percent after a llama joined their flocks. And 80 percent rated their llamas as effective or very effective for guarding sheep. In another study conducted in Utah, 90 percent of producers rated guardian llamas either effective or very effective on the job. National Agricultural Statistics Service figures indicate that 14 percent of sheep and goat producers had llamas on guard in 2004.

Llamas naturally dislike dogs and coyotes and tend to make excellent guardians, although there are definite exceptions to the rule. Llamas prefer the company of other lamas, so it's wise to use just one llama per pen or herd of hoofed stock; however, multiples work fine for protecting alpacas. Intact males are often aggressive toward hoofed stock and shouldn't be used.

Llamas require the same food, vaccinations, and hoof care that sheep and goats do, so it's easy to treat them as just another member of the herd. Not so with donkeys (even a small feed of Rumensin-laced feed can kill a donkey) or livestock guardian dogs.

On the other side of the coin, most llamas are more aloof than are donkeys or guardian dogs. Where most donkeys and dogs crave human interaction, the average llama doesn't. This makes it more difficult to catch and handle some llamas for routine maintenance chores. And, llamas don't do as well as donkeys or guardian livestock dogs in hot, muggy climates, where they're prone to heat exhaustion. Depending on the length of their fiber, some llamas require full or partial shearing at least once a year.

It's also important to realize that llamas are prey animals, too, and relatively defenseless; they don't stand a chance against a pack of dogs or a mountain lion or bear. Guardian llamas make sense where a neighbor's dog might worry livestock or a hungry fox or coyote might want to dine on newborn lamb. Don't expect llamas to take on the big guns and survive. Many guard llamas do their best to save their charges and are mutilated or killed for trying.

Sell or Spin Your Lama's Fiber

People have been shearing lamas and using their fiber for thousands of years—you can, too. Lama fiber isn't true wool. One of the beauties of lama fiber is that its physical structure is more like hair, yet its softness and fineness enable the handspinner to produce wonderful yarn with ease. Lama fiber is semihollow as well, which makes it a wonderful insulator (alpaca fiber is seven times warmer than sheep's wool). It's easier to process because lama fiber lacks the lanolin that makes unwashed sheep's wool feel greasy to the touch.

Advice from the Farm

Sounding the Alarm

The experts discuss guardian llamas.

Saviors of the Day

"There are many stories about great guardian llamas saving the day for their flocks. However, llamas are no match for some predators, and in particular for packs of dogs. Llamas should be seen as one part of the herd's defense against predators. Don't depend on only your llama for protection."

—*Tina Cochran*

The Intimidators

"Either a female or a male can be an effective guardian animal. Some feel that a single animal will be more effective, but there seem to be just as many stories of llamas working very well in pairs. Mine all rush to defend by intimidation, but only one will alarm call. I think that's because they are in there with Anatolian Shepherds and they pretty much depend on the dogs to do the work, although they all rush over to see what's up.

"Many alpacas exhibit guardian tendencies, but their smaller stature doesn't intimidate the way a llama can."

—*Deb Logan*

Really Good Fences

"Alex is a great guardian llama, but we don't expect him to do it all by himself. We have good—and I mean really good—fences, and we shut the sheep and Alex in a secure pen near the house at night or if no one is home. Also, some of our sheep wear collars with bells on them so we know if anything is chasing them."

—*Jan Johnson*

A Winning Combination

"Llamas can be very effective guardians in combination with livestock guard dogs. A llama will unfailingly spot anything unusual in the pasture, and it will either venture forth with the LGD [livestock guard dog] or attempt to move smaller stock back to a protected area while the LGD addresses the transgressor.

"It's important to note that a llama by itself is not an effective guardian. Llamas are more like sentries who attempt to intimidate and in some cases stomp smaller predators such as coyotes, foxes, or the odd dog. They are ineffective against packs of dogs or coyotes, big cats, bears, and the like. Dog packs are the biggest killer of llamas."

—*Deb Logan*

This fiber from one of Golden Heartland Alpacas' studs, Curly Eye's Oberon, is a fine example of striking color and crimp.

Sheep's wool requires careful washing prior to use; comparatively speaking, cleaning lama fiber is a snap.

The Internet is a treasure trove of information about lama wool: its properties, qualities, and how to shear it, process it, and felt the fiber or spin it into yarn. To begin, visit your favorite search engine and run a search using the key words *llama fiber* or *alpaca fiber*. Don't overlook the wide array of fiber articles archived at the *Alpaca World* and *International Camelid Quarterly* magazine Web sites (see Resources)! For up-to-date spinning instructions, search using the keywords *spinning alpaca* fiber or *spinning llama fiber* (note: alpaca resources apply to spinning dehaired llama fiber, too); for *felting* instructions, substitute felting for *spinning* when you run your search.

CHAPTER NINE

Making Money with Llamas and Alpacas

Can you make good money with hobby farm lamas? The answer is a qualified maybe.

Breeding Stock for Sale

It's certainly possible to profit from selling breeding stock, but it isn't necessarily easy. If you think you're up for the challenges, here are some preparations you need to make and points you need to consider.

Essential Steps

Start with high-quality, registered stock. You'll show more profit selling crias from two outstanding females than you will from six average ones or a whole slew of unregistered females from the sale barn. Start small if you must and expand your operation by retaining your best female crias for your breeding herd.

Set up your farm as a credible breeding operation. It needn't be fancy, but facilities must be maintained in a safe and tidy manner. If you plan to stand a male at stud to visiting females, provide safe, separate accommodations for them.

Don't breed solely for short-term market fads such as popular bloodlines, fiber type, or color, but do take them into consideration. Quality first, *then* the frosting on the cake: that should be every breeder's goal.

Work with your veterinarian to establish worming and vaccination schedules tailored for your herd and your locale. Keep those toenails trimmed. Maintain your animals in tip-top condition.

Bandit's ears are shapely but they point too far forward, giving him an odd appearance, despite his otherwise fine conformation.

Bandit's ears look even stranger from the front. Their unconventional shape means Bandit should not be used for breeding stock.

Keep careful health, breeding, and training records, and be ready to provide copies to buyers. Likewise, register crias as soon as they're born and have their papers in hand when buyers come looking.

Don't breed young females until they're physically and mentally mature. Geld most of your young males, keeping only the cream of each year's crop of males intact. *Never* sell a poor-quality male as breeding stock, especially if he carries your farm prefix. If you are unsure about an individual, geld him anyway or keep him until you decide for certain.

Be totally ethical at all times. Represent your animals in a positive light but point out major failings, especially when selling to buyers sight unseen. It's good for business and simply the right thing to do.

Stand behind the lamas you sell. Use a sales contract or written guarantee for every sale or service. Offer after-sales support. Satisfied customers are your best advertising.

When you sell a llama or alpaca, send it home wearing a new halter that fits, a new lead rope, and a folder of written information about its accustomed diet;

Llama ears should be long and shapely and point in at the tips. Properly placed ears are called banana ears.

complete health records (and breeding records if applicable); a photocopy of the lama's registration certificate; and a copy of its extended pedigree. When selling homebred crias, tuck in an 8-by-10-inch photo of each parent.

Mail the transfer to the registry yourself (this is why you've provided a copy of the registration certificate instead of the original); buyers often forget to do it and over time, papers can be lost. Ask your registry to mail transferred papers directly to the buyer and make sure he understands this policy.

Business Contracts 101

When you sell a llama or an alpaca, use a legal sales contract and use it every time. Before you sell your first animal, hire a lawyer to draw up a contract specifically for your farm and its needs. It should:

1. Indicate the date of sale.
2. Identify both parties. That's you (the seller) and the buyer, including full names, addresses, phone numbers, and Social Security or federal tax identification numbers.
3. Identify the llama or alpaca. Include its name, age, color, markings, registration number, and the names and registration numbers of its sire and dam.
4. Set forth the agreed-upon price and terms of sale. If the buyer pays a deposit, when is the balance due, and if the buyer backs out, will the deposit be returned or forfeited? If there are special contingencies to the sale, spell them out in detail. When selling on installments, indicate the interest rate, payment schedule, and who will retain possession of the lama and its registration papers until the transaction is completed. Indicate what will happen if the buyer misses payments or defaults.
5. Spell out additional costs. How long you will agist (board) the llama or alpaca at no additional cost? Should

Did You Know?

In 2002, the CamelidMed division of the College of Veterinary Medicine at Ohio State University mailed out an extensive questionnaire to determine the prevalence of lameness in camelids. These are some of their findings.

- A total of eighty farms responded. Surveys were returned from 18 states, as well as Canada, Australia, and New Zealand. Of the farms reporting, 89 percent were breeding operations, 68 percent produced fiber for textiles, 64 percent showed their lamas, 63 percent considered their lamas pets, and 26 percent used them for other purposes, such as packing, therapy work, driving, commercial trekking, and flock guardians.
- A total of 2,172 camelids were included in the study with an average farm size of 27 animals. Of the 2,172 animals, approximately 54 percent were llamas, 38 percent were huacaya alpacas, 8 percent were suri alpacas, and one was a guanaco.
- The average farm size was 115 acres.

Advice from the Farm

Shear Madness

The experts discuss grooming and shearing.

Get the Dead Wool Out

"We groom our llamas with various sizes of soft pin brushes as well as a regular pin-cushion dog brush. We also have rug beater looking things that you can use to separate and fluff out various junk, but if the animal is at all matted, you can forget that.

"I like to take about ten minutes each evening and lightly groom the llamas' necks—all of mine seem to enjoy it, and it's sort of a bonding ritual.

"Many owners obtain used street sweeper brushes from their municipality and mount them either vertically or horizontally so the llamas can brush up against them. They seem to enjoy it, and it really helps get the dead wool out."

—Deb Logan

Once a Year—Just Do It!

"Our alpacas are sheared once a year using electric shears. We hire a shearer from a nearby larger alpaca farm to do the shearing. Alpaca fleece is a valuable commodity, and paying for shearing saves us more money than we spend by not having second cuts and knowing that our alpacas are taken care of properly.

"Shearing day is the biggest day in the alpaca year, so hire an experienced person to do it, invite everyone you know to come and watch, and make it a great day rather than a stressful time for you and your animals.

"A tip: whether you use electric shears, clippers, or hand shears and whether you hire someone or do it yourself, just DO IT! An unshorn alpaca is an uncomfortable and unhealthy alpaca and potentially even a dead one."

—Tina Cochran

No Naked Baby Hamsters

"We use Premier 3000 shears and Premier 3000 clippers with changeable shearing heads to shear our llamas. That way we can set up two pairs of shears or keep one for use as a clipper. What we use depends on the desired end result.

"Lately we've been using shears with a camelid blade and a lift bar on them; this leaves a bit more wool on the animal. They look very uniform, but it takes a bit of practice to get the hang of it to ensure a smooth cut. It's especially great for light-colored animals with pink skin, as this way they aren't as prone to sunburn, plus they don't end up with that 'naked baby hamster' look.

"The good thing about electric shears is that they're fast, and you can shear a lot of animals in a day. On the other hand, shears can easily cut an animal, and if you don't pay attention, they can get hot and burn the skin. Durable clippers and shears are typically expensive, and you must have an adequate number of blades. If you don't have a means to blow dirt and debris from the fleece before you begin, then you'll use more blades. Lastly, they require electricity. If you don't have electricity in your barn, then the animal has to be moved elsewhere for shearing.

"If we need a show cut, sometimes we use the shears and then go back over certain areas with a medium blade in the clippers for a closer cut.

"We have also used Fiskars scissors on certain animals. The good and the bad, depending on your perspective: a scissored finish is definitely not as smooth (think divots on a golf course), but after a month or so it all evens out. This approach takes longer, but it's often a lot less stressful to both the shearer and the designated victim! But it could take forever to do a large herd."

—Deb Logan

the buyer collect the animal at your farm, or will you deliver? If so, at what price? If the lama requires vet work prior to shipment, who pays?

6. Address attendant risks. Indicate the date when the buyer will assume full responsibility for the lama's illness, injury, or death or for damage or injuries inflicted by the lama while in your care.
7. Describe applicable guarantees. Spell out the specifics in detail.
8. If applicable, indicate the need for insurance. When selling lamas on time payments, insist on insurance at least to the extent of the unpaid balance, at the buyer's expense, with yourself named as the loss payee.
9. Incorporate the signatures of every party named in the contract.

More Great Ways to Earn Money with Lamas

You won't get rich engaged in the following enterprises, but they're unusual and interesting enough to attract attention; if you run your business well, you're likely to succeed. Try one or more on for size. If one fits, go for it!

Market Fiber or Products Created with Lama Fiber

A great way to support your lamas is by marketing their fiber as roving (clean and carded material ready to spin) or yarn (spin it yourself or hire someone). Or you can offer finished articles such as knitted mittens and scarves or felted slippers. You needn't open a full-fledged

This South American chulla hat is crafted of soft, cushy alpaca fiber. Also pictured are alpaca fibers in several colors (on the earflap) and white ccara llama underfiber.

store: llama and alpaca roving and yarns sell well on eBay.

Or place an ad in your local paper and offer to shear llamas in exchange for their wool (be sure to stipulate that only reasonably mat-free llamas are eligible). Then, process the wool, and sell it through your online store or at craft and fiber gatherings.

Market Lama Poop

Yes, lama droppings. Call it llama beans or paca poo, there are buyers for the fruits of your dung piles.

Lama manure is one of the best organic fertilizers for vegetable and flower gardens, lawns, and houseplants. When dry it's practically odor free, and compared with other livestock manures, it's brimming with nutrients, such as nitrogen, phosphorus, and potassium (elements that form the familiar N-P-K ratio printed on fertilizer bags). Due to camelids' efficient digestive process, lama manure contains much less organic matter than most other manures do. When organic matter decomposes, it creates heat, and this is what burns tender seedlings when the fresh manure of other species is applied to plants. Lama manure works fresh as long as it isn't soaked with urine, but dried is better and composted is best. That gives lama poo entrepreneurs plenty of ways to market their product.

For instance, one good way to use lama beans is as manure tea. To make this liquid fertilizer, place 1 cup of lama dung in a capped 1-gallon container of water and let it set for three days. Once it's set, shake the container and use the liquid to water your plants.

Did You Know?

- The current South American lama population is estimated to be about 7 million head.
- Peru's alpaca population numbers about 3 million. Bolivia has approximately 500,000 alpacas, while Chile and Argentina have about 50,000 alpacas between them.
- The International Alpaca Association (the Asociacion Internacional de la Alpaca/AIA), located in Arequipa, Peru, reports that the country produces 4,000 tons of alpaca fiber annually. Boliva produces the largest amount of llama fiber at 600 tons.

Livestock Manure Comparison

Animal	Nitrogen %	Phosphorus %	Potassium %
Llama/Alpaca	1.7	0.69	0.66
Chicken	1.0	0.8	0.4
Cow	0.6	0.15	0.45
Goat	2.0	0.5	0.6
Horse	0.7	0.25	0.55
Pig	0.5	0.35	0.4
Sheep	0.95	0.35	1.0

Or customers can mix 1 cup of beans to two quarters of potting soil as an organic enhancement. They can spread dry or composted manure over their gardens 3 inches deep, then rototill it in before planting. Or users can feed trees by piling two inches of dung around their trunks before mulching for the season. The applications are endless!

Most folks who sell lama manure do it in one of several ways. They spread it out and allow it to air dry, or they compost it, then package the results in manageable sizes. The sellers deliver it by the pickup truckload, or they allow buyers to come to the farm and load it themselves.

Consider selling nicely labeled ziplock bags (sandwich to 1-gallon size) of lama manure through your farmers' market store, online, or wholesale to garden centers. Placing ads in local pennysavers to sell larger amounts; or take the no-frills approach by shoveling it into used feed bags and setting it by the road gate with a sign and honor-system cash jar. ("Llama manure—$3 a bagful!")

Man a Shearing and Toenail-Trimming Service

No one looks forward to shearing and nail-trimming time, especially people who don't own the equipment needed to keep everything running smoothly. The enterprising lama entrepreneur who buys a portable chute and top-of-the-line shears and clippers, then learns to apply them properly, will have more business than he or she can handle. Trust me.

Lama Farm Sit

Llama and alpaca owners often can't travel because they're loath to leave their animals in the hands of generic house or pet sitters. A knowledgeable person with lama experience, who advertises services in lama publications and regional livestock periodicals, will likely find all the work he or she needs to stay busy year-round.

Offer Llama Trips

When most of us think of llama packing, we visualize long pack trips into the Western mountains. However, no matter where you live, you can offer day hikes and overnight llama camping. Add the joy of llamas to your own personal arena of expertise, run your business well, and customers will flock to your door. Food for thought: offer overnight trips into the woods at your favorite state park, serve old-fashioned suppers of savory beef stew and homemade biscuits, and tell regional folk tales around the campfire. Lead day hikes for families or women's groups; serve hot dogs and beans or even gourmet fare. Teach would-be wild foodies and native herbs enthusiasts to identify, harvest, and prepare the wild things growing in your neck of the woods. Pack llamas make every trip into nature more fun.

Become a Lama Whisperer

The vast majority of lama breeders would gladly pay a patient, truly knowledgeable person to teach their crias to lead, load, and pick up their feet. The same goes for skittish, older llamas and alpacas, too. Practice your technique; buy a strong, portable catch pen and a truck to haul it on; and advertise. The work is out there in spades.

John carefully desensitizes Bandit's legs before trimming his toenails. Experienced trimmers can market their services for profit.

Takin' Care of Business

No matter what direction your lama enterprise takes, here are some easy, low-cost steps you can follow to boost sales and improve your chance of success.

Choose a Memorable Business Name

Pick a business name people can remember, spell, and pronounce. Don't appropriate a name another lama entrepreneur is already using. To prevent unpleasant litigation later on, it never hurts to find out whether the name you select is trademarked. If it isn't, consider having it trademarked yourself. To learn how to trademark your farm or business name, visit the United States Patent and Trademark Office Web site at www.uspto.gov/go/tac/doc/basic.

"Lama Whisperer" John makes friends with Bandit, who was a relatively unhandled, uncastrated male when we got him.

Promote!

If people have never heard of you, they can't buy your product or avail themselves of your services. Here are some ways to help buyers find you.

Go Online

With the help of basic Web authoring software or a good book about Web site design, every business owner can build the kind of site that sells—I guarantee it!

Spring for your own domain name. Don't use freebie-hosting services replete with pop-ups and advertising banners on every page. A short, punchy URL (Web address) is more effective than a long one. Compare www.myfreewebservice.com/~GreatGuanacos.html with www.GreatGuanacos.com. Which would you remember? When choosing a domain name, farm or business names work best. If your farm name is Bodacious Llamas, try for BodaciousLlamas.com. If that's taken, opt for BodaciousLlamas.net, -.biz, -.info, or -.us. Or dream up a catchy phrase such as Llamas4U, Llama-Llama-Llama, WorldsBestLlamas, or FantasticFiberAlpacas to use instead of your farm name (all of these are available as .com URLs as we go to press).

It's in the Cards

Business cards are a budget-wise promoter's best friend. Never leave home without them.

- Tack cards to every bulletin board you pass. Use sturdy pushpins so you can stack them; this encourages interested parties to take one along.
- Ask businesses to display your cards near their cash registers. Target veterinarians and farm stores in your area.
- Craft your own gift tags on the backs of your business cards.
- Hand cards to people you meet. Ask them to take several and pass the extras along to their family and friends.
- Buy a conference-style name tag holder, insert your card, and wear it on your lapel; it's a terrific conversation starter!

Lamas Beyond the Sidewalk

The best way I know of to learn the nuts and bolts of country entrepreneurship is to read rural business guru Ellie Winslow's books (*Marketing Farm Products: And How to Thrive Beyond the Sidewalk* and *Growing Your Rural Business: From the Inside Out*) or to attend one of her alpaca marketing workshops. Watch for them in conjunction with alpaca shows and gatherings nationwide or visit her Beyond the Sidewalk Web site (see Resources) for information.

It's tempting to settle for freebie cards you can order online, but don't. The card company's advertising on the back of freebie cards distracts from your image. *Cheap* isn't a word you want associated with your enterprise. Go the extra mile, and pay for top-flight cards.

If you spring for quality cardstock (not perforated punch-outs) and you understand layout design, you can probably print your own. Some word processing software and most desktop publishing programs offer create-a-card capability. If you can do a professional-looking job, do it; if not, order business cards printed by the pros.

Opt for a simple design. Avoid hard-to-read fonts such as script, and use no more than two font families per card. Your name or farm name should be the largest text element on the card. Consider having a custom logo made for your business printing; barring that, clip art or a good photo is a nice touch. (A poor photo is worse than none at all, so choose a flattering pose, and resolution should be spot-on.)

Run a spell-checker over your finished card before you print—never omit this important step. Nothing says "amateur" more emphatically than advertising replete with a slew of misspelled words. Keep your card up to date; don't scratch through or white out text and write in corrections—get new cards.

And don't distribute bent or soiled cards that reflect poorly on your business. Take care of your business cards. Keep them crisp and clean in attractive business card holders or tucked in their own section of your wallet.

Pennywise Promotion

Effective promotion needn't break the bank. Invest in truck and trailer lettering. Incorporate your phone number, e-mail, and Web site address into the design. Turn nonbusiness vehicles into rolling billboards by affixing custom-designed magnetic signs.

Opt for custom-printed checks, invoices, and other business forms imprinted with your logo and business information. Buy T-shirts, jackets, and hats printed or embroidered with your logo and farm name, and wear them everywhere you go. Order extras for your customers and friends.

Erect a large, legible road sign by your driveway. If you live on a well-traveled road, display lamas in a pasture adjacent to the highway. If your females have crias, put them by the road; nothing stops traffic better than adorable baby animals.

Take your animals out in public. Show. Train a driving or pack llama, and drive or lead it in parades with farm signs attached to the cart or panniers. Take a friendly llama or alpaca to hospitals and nursing homes to visit shut-ins; have fun, gain exposure, and make people happy—everyone wins! Take llamas or alpacas and a display booth to community gatherings, county fairs, and farm expos.

Give talks and demonstrations. Many civic groups need speakers for meetings and events; let it be known that you're available and interested. Hold an open house. Sponsor and lead a 4-H pack llama or lama fiber project. Host a shearing demonstration, training clinic, or lama business seminar at your farm.

Read lama stories to Head Start or kindergarten children or volunteer to read them at your library's story hour. (*Llama Llama Red Pajama* and *Llama Llama Mad at Mama*, both written by Anna Dewdney, and *Stop Spitting at Your Brother*, by Diane White-Crane, are almost certain to be hits.) Award plush toy llamas or alpacas to the children who dream up the cutest names for a cute and cuddly cria you take to their school. (But be sure to distribute nice consolation prizes to the rest of the class.) Tip off your local newspaper in advance.

Participate at llama- and alpaca-related Listservs and e-mail lists; be the first to field questions. This establishes you as a knowledgeable party and costs nothing but your time. People buy from people they know and respect, and you can advertise on many lists for free.

Add a business-related signature to your outgoing e-mail; incorporate your farm name, location, a tagline describing your services, and your farm's Web address.

Use your imagination, and promote your business for pennies!

This mosaic in the drive at Klein Himmel Llamas symbolizes the farm's firm commitment to 4-H youth in the community.

Acknowledgments

Thanks again to the good folks who contributed "Advice from the Farm" tips and words of wisdom.

Tina Cochran operates Golden Heartland, the "Exclusive Alpaca Boy's Club" in Adrian, Missouri. The Cochran family specializes in helping people get a start in alpacas on a tight budget. Contact Tina at http://www.goldenheartland.com or send an e-mail to tina@goldenheartland.com.

Mary Collins, an avid handspinner and rug weaver, lives near Batesville, Arkansas, with her fiber alpaca buddies, Cash and Cary.

Nancy Frank is a lifelong animal enthusiast and trainer, twenty-year owner of Opportunity Llamas, and current director of Companion PAWS (Pets and Wellness for Seniors at http://www.companionpaws.net), an organization that works with animals to enhance the lives of seniors.

Bob Huss is a kind soul who loves llamas, especially his girls Betsy and Muffin, both of whom are special needs llamas. Betsy, a nervous survivor of abuse, and Muffin, a recovering ABS llama, love clicker training and Bob's famous homemade llama bars.

Jan Johnson lives near Anoka, Minnesota, on a hobby farm where she raises Cheviot sheep that are ably guarded by Alexander the Great, a beautiful classic llama.

Deb Logan is the Georgia coordinator for Southeast Llama Rescue. She resides at Wit's End Farm in Acworth, Georgia, in the company of a passel of happy llamas.

And a special thank you to Glen and Margo Unzicker of Klein Himmel Llamas (http://www.kleinhimmel.com) in Goshen, Indiana, for allowing John to photograph their gorgeous llamas.

Appendix: Lama Maladies at a Glance

Most llamas and alpacas are remarkably healthy, yet there are diseases and abnormalities you should watch out for. These are some of the most important.

Bovine Viral Diarrhea Virus

Bovine viral diarrhea virus (also called BVD or BVDV) is a growing concern among lama owners. Acute infections are generally mild, but if a female is infected during pregnancy, she could abort or her cria could suffer congenital birth defects. Or the cria could be born with a persistent infection; then it will shed huge amounts of the virus throughout its lifetime, potentially infecting many others. Not a lot is known about the disease in camelids, but due to its potentially devastating effect on the burgeoning alpaca industry, research is an ongoing priority.

Brucellosis

Brucellosis is a serious, federally reportable disease also called Bang's disease in cattle, brucella ovis in sheep, and undulant fever in humans. Brucellosis is caused by bacteria from the genus *Brucella*. Brucellosis (as undulant fever) can be passed to humans who handle affected aborted fetuses. Fortunately, brucellosis is rare among lamas. There is a blood test for brucellosis (and the test is required for entry into some states), but the test sometimes gives false readings. Therefore, any lama that tests positive for brucellosis should automatically be retested.

Symptoms: Spontaneous abortions, retained placentas; intermittent fevers; and stiff, swollen joints.

Treatment: With few exceptions, state and federal laws require the slaughter of affected animals.

Prevention: Effective brucellosis vaccine is available, but since it's off-label for lamas, consult with your veterinarian before using it.

Caseous Lymphadenitis

Caseous lymphadenitis (commonly referred to as CL and also called cheesy gland) is caused by the bacterium *Corynebacterium pseudotuberculosis*. It's not a common problem with lamas; however, lumps on lamas kept with infected sheep and goats should always be suspect. CL manifests as thick-walled, cool-to-the-touch lumps containing odorless, greenish-white, cheesy-textured pus. CL abscesses form on lymph nodes and lymphoid tissue, particularly on the neck, chest, and flanks, but also internally on the spinal cord and in the lungs, liver, abdominal cavity, kidneys, spleen, and brain. Transmission is via pus from ruptured abscesses.

Symptoms: Lamas with internal abscesses may fail to thrive, depending on which organs are involved.

Treatment: Any lama with a ripening abscess should be quarantined. The abscess should be drained and treated according to your veterinarian's instructions. Don't allow drained pus to contaminate your farm. Because CL is transmissible to humans, it's important to wear protective clothing. When the procedure is completed, sterilize everything you used or wore, or burn it along with anything else contaminated by pus. The lama should remain quarantined until its abscess has fully healed.

Coccidiosis

See chapter six.

E-Mac

E-Mac is the common name for infestation by *Eimeria macusaniensis*, a serious and unfortunately common type of coccidian, the same type of organism that causes everyday coccidiosis. Found in 28 percent of all Midwestern alpaca herds in 1999, E-Mac can kill lamas of all ages but is especially prevalent in adults.

Symptoms: Lethargy, depression, weight loss, failure to thrive in young lamas. Unlike common coccidiosis, diarrhea is transient or rare.

Treatment: Deworming medications have no effect on E-Mac. E-Mac is usually treated with the same drugs used to treat common coccidiosis: sulfa drugs and amprolium.

Prevention: Weekly manure pickup is essential. Avoid feeding lamas off the ground. Don't house immunosuppressed lamas in areas where crias are kept, since infested crias tend to shed more oocysts than adult lamas do.

Enterotoxemia

Enterotoxemia, also known as entero, overeating disease, and pulpy kidney, is caused by common bacteria found in manure, soil, and even the rumens of perfectly healthy lamas. Although enterotoxemia is relatively rare in lamas, overeating on grain or milk and abrupt changes in quantity or type of feed can cause bacteria to quickly and dramatically proliferate;

these bacteria produce toxins that can kill a lama in hours.

Symptoms: Bloating, standing in a rocking horse stance, teeth grinding, seizures, foaming at the mouth, coma leading to death.

Treatment: Treatment is usually ineffective because death occurs so quickly.

Prevention: The most common vaccine is CD/T toxoid (T is for tetanus; this three-way shot also prevents that potentially fatal disease); it's also available in an eight-way vaccine sold as Covexin 8. CD antitoxin provides short-term immunity to previously unvaccinated lamas.

Facial Swellings

Facial swellings are fairly common among lamas and can be caused by a number of things. Before assuming your lama has a facial swelling, check to make certain it isn't simply holding a cud (a food bolus) inside one cheek. Press the swelling; if it's soft and it disappears after a while, it was cud. Caseous lymphadenitis abscesses (see previous page) often occur on the lymph nodes of the neck; if you suspect CL, consult your vet before the center of the abscess softens and bursts. Abscesses also form when any of hundreds of organisms breach the skin via puncture wounds, splinters, and everyday cuts or abrasions.

Hard, bony masses along the jaw or on the face may be bone infections or tooth root abscesses; these are difficult to diagnose and treat, so again, consult your veterinarian.

Heat Stress

Heat stress is a major killer of lamas, particularly in the Southern states where both heat and humidity factor into the equation. Lamas evolved in the high Altiplano of the South American Andes, so they aren't adapted to long periods of extreme heat and humidity. According to the International Lama Association's educational brochure, *Heat Stress in Llamas*, you can add together the ambient temperature (degrees Fahrenheit) and humidity to help determine if your lamas are at risk on any given day. If this number is less than 120, little risk exists; if the number is 150 or more, as many precautions as are available should be implemented; as the number approaches or exceeds 180, lamas are at great risk for developing heat stress. Old, young, sick, high-strung, and obese individuals are particularly prone to heat stress.

Symptoms: Panting; depression; not eating; a rectal temperature of more than 104°F and a heart rate of more than 90 beats per minute; drooping of the lower lip, facial paralysis, slobbering; trembling, weakness; scrotal swelling in intact males; collapse; seizures; death.

Treatment: Cool the lama down; hose its underside with cool water (if the lama has fiber and you hose its body, be certain to soak it all the way to the skin). Place the lama in the shade or in a pond or swimming pool, in front of fans, or in an air-conditioned room. Call your vet!

Prevention: Shear lamas in the spring (even show lamas should be barrel clipped). Don't let lamas get fat. Don't breed females so they give birth during the summer months, and avoid weaning crias during the same period. Provide shade and plenty of cool (but not ice cold) drinking water. Consider furnishing lamas with children's wading pools or wet sandboxes (many lamas love to kush in them), fans, or sprinklers. Avoid moving lamas to a warmer climate during the summer months.

Johne's Disease

Johne's disease (pronounced YO-nees), also called paratuberculosis, is a contagious, progressively fatal, slow-developing disease of ruminants. It's most commonly seen in dairy cattle. Johne's is caused by *Mycobacterium paratuberculosis*, a close relative of the bacterium that causes tuberculosis in humans, cattle, and birds. The disease is rampant worldwide; according to Johne's Information Center statistics, 7.8 percent of America's beef herds and 22 percent of our dairy herds are infected with *M. paratuberculosis*. Johne's disease typically enters lama herds when an infected but healthy-looking animal is added to the mix. The infection then spreads to its herd mates. (Lamas, generally crias, become infected through oral contact with contaminated manure from an infected lama.) Crias are also infected by nursing from infected dams.

Symptoms: Progressive loss of condition; weakness.

Treatment: None.

Listeriosis

Listeriosis is an uncommon but serious disease caused by a bacterium called *Listeria monocytogenes*. It's found in soil, plant litter, and water and even in healthy lamas' guts. It is a type of encephalitis (inflammation of the brain). Problems arise when dramatic changes in feed or weather conditions occur, causing bacteria in the gut to multiply. Parasitism and advanced pregnancy can trigger bacteria proliferation, too.

Symptoms: Disorientation, depression; stargazing, staggering, weaving, circling; one-sided facial paralysis, drooling; rigid neck with head pulled back toward flank. Symptoms resemble polioencephalomalacia, rabies, and tetanus.

Treatment: Treat according to your vet's recommendations.

Prevention: Avoid drastic changes in type and amount of feed, and never feed moldy hay or grain.

Meningeal Worm

See chapter six.

Mycoplasma

Mycoplasma was previously known as eperythrozoonosis or EPI, but its name has been changed to reflect the classification of the bacterium that causes it (from *Eperythrozoonosis suis* to *Mycoplasma haemolamae*). The bacterium attaches itself to the red blood cells of lamas. Studies indicate that 25 percent of camelids in the United States are infected. Some infected lamas appear healthy but are carriers; others, particularly individuals with depressed immune systems, manifest symptoms. Researchers believe the disease is spread via infected blood (think biting insects, reused needles, and blood transfusions).

Symptoms: Anemia; chronic weight loss; depression, lethargy; staggering, stiffness of the hindquarters; collapse.

Treatment: Oxytetracycline can control the infection, but "cured" animals often become carriers.

Polioencephalomalacia

Polioencephalomalacia (also called PEM, or cerebrocortical necrosis) isn't related to the viral disease called polio (poliomyelitis) in humans. Polioencephalomalacia is a neurological disease caused by a thiamine (B_1) deficiency that culminates in brain swelling and the death of brain tissue.

Symptoms: Disorientation, depression; stargazing, staggering, weaving, circling, tremors; diarrhea; apparent blindness; convulsions; death.

Treatment: If treatment begins early enough, affected lamas given thiamine injections begin improving in as little as a few hours.

Prevention: Thiamine deficiencies can be triggered by eating moldy hay or grain; overdosing with amprolium (Corid) when treating for coccidiosis; ingestion of certain toxic plants; reactions to dewormers; and sudden changes in diet, including weaning. Overuse of antibiotics contributes to thiamine deficiencies.

Stomach Ulcers

Stomach ulcers are surprisingly common in lamas and they can be fatal, so it's important to recognize and prevent this disease. Most ulcers are triggered by stress. Newly weaned crias, females whose crias have recently been weaned, and lamas transported into new surroundings away from familiar faces are particularly prone to stress-induced ulcers. Diets high in concentrates can contribute to ulcer formation. An ulcer occurs when overproduction of gastric acid upsets the pH balance in C-3, the third compartment of a lama's stomach. If the ulcer penetrates the wall of C-3, stomach contents spill into the abdominal cavity causing peritonitis, and the lama dies.

Symptoms: Teeth grinding (an indication of pain), pained facial expression; depression; kushing in abnormal positions, kicking at belly, excessive rolling; not eating; black stools indicating internal bleeding.

Treatment: Remove the offending stress, and consult your veterinarian for treatment options.

Tetanus

Tetanus occurs when wounds are infected by the bacterium *Clostridium tetani*. These bacteria thrive in anaerobic (airless) conditions such as found in deep puncture wounds, fresh umbilical cords, and wounds caused by recent castration. Unless treated early and aggressively, tetanus is nearly always fatal.

Symptoms: *Early on:* stiff gait; mild bloat; anxiety. *Later:* standing in a rigid rocking horse stance; drooling; inability to open the mouth (hence tetanus's common name: lockjaw); tail and ear rigidity; seizures; then death.

Treatment: If you suspect tetanus, contact your veterinarian without delay.

Prevention: All lamas should be vaccinated to prevent a horrible death.

Urinary Calculi

Urinary calculi (also referred to as UC or urolithiasis) are mineral salt crystals (stones) that form in the urinary tract and block the urethras of male lamas. (Both sexes are afflicted, but stones pass easily through females' relatively larger, straighter, and shorter urethras.) The condition requires immediate medical attention. UC doesn't correct itself, and if left untreated, the afflicted animal's bladder will burst and the animal will die.

Symptoms: Anxiety, restlessness; teeth grinding, pained facial expression; straining to urinate; standing in either a rocking horse or hunched over stance; impaired flow of urine (dribbling).

Treatment: Surgical removal of offending stones is generally necessary. Consult your veterinarian if you suspect UC. Early intervention makes the difference between recovery and certain death.

Prevention: It's important to feed male lamas a balanced 2:1 calcium-phosphorus ration. Adding minute quantities of ammonium chloride to the diet may prevent some types of calculi from occurring. Supply male lamas with plentiful supplies of clean, palatable drinking water; encourage water consumption by adding a heater in the winter and keeping drinking water in the shade during the summer.

White Muscle Disease

White Muscle Disease, also called nutritional muscular dystrophy, is caused by a serious deficiency of the trace mineral selenium. Most of the land east of the Mississippi and much of the Pacific Northwest is selenium-deficient; these are the areas where white muscle disease is most likely to occur. The best source of information about selenium conditions in your locale (especially in areas where selenium levels may vary widely from farm to farm) is your county extension agent.

Symptoms: *Crias:* stillbirth; weakness, inability to stand or suckle; tremors, stiff joints, neurological problems. *Adults:* infertility; abortion, dystocia, retained placentas; stiffness, weakness, lethargy.

Treatment: Make certain you are located in a selenium-deficient area before treating for white muscle disease. Injections of BoSe (a prescription selenium and vitamin D supplement) sometimes dramatically reverse symptoms, especially in neonatal crias.

Prevention: All lamas raised in selenium-deficient areas should be fed selenium-fortified feeds, have free access to selenium-added minerals, or be given BoSe shots under a veterinarian's direction. To prevent birthing problems and protect unborn crias, females should be injected with BoSe four to six weeks prior to giving birth (consult your vet for particulars).

Glossary

Aberrant behavior syndrome (ABS)—a condition in which a lama (usually an uncastrated male) that was improperly imprinted on humans while it was a cria becomes dangerously aggressive toward people as an adult. Previously called berserk male syndrome (BMS).

Accoyo—breeders' prefix of the famous alpaca breeder Don Julio Barreda of Peru; imported animals that originated on his ranch carry this name.

Agistment—an arrangement by which a person boards his animals at an establishment other than his own.

Alarm call—the sound a llama makes when it feels the herd is threatened. Some say it resembles the call of a tropical bird; others, that it sounds like a turkey call.

Altiplano (al-tee-PLAH-noh)—the high plateau region of southern Peru and northern Bolivia.

Angular limb deformity—the deviation of a limb, either out- or inward; can be congenital or acquired.

AOBA—Alpaca Owners and Breeders Association.

Appaloosa—a multicolored lama usually having dark spots on a lighter background of tan, brown, or white.

Apron—longer, more highly medullated fiber on a lama's chest.

ARI—Alpaca Registry Incorporated.

Aymara (ay-MAR-ah)—aboriginal people of the Andean and Altiplano regions of Peru, Bolivia, northern Chile, and northeastern Argentina, who speak the Aymaran language.

Banana ears—banana-shaped llama ears that come up and then curve inward toward one another.

Berserk male syndrome (BMS)—*see* Aberrant behavior syndrome.

Bib—*see* apron.

Blanket—the best part of a lama fleece, beginning at the shoulder and running the full length of the back to the base of the tail and then down a little more than halfway past the center of the sides; all of a fleece except fiber from the legs, chest (apron), belly, and the britch. Approximately 60 percent of the total fleece.

Body score—a numerical value given to an animal based on how fat or thin it is. Some systems are based on scores of one to five (with one being emaciated and five obese), others on scores of one to ten.

Bone—a lama's skeletal size based on measurement of the bone circumference of the lower legs; a lama with "a lot of bone" has a large frame and sturdy legs.

Bonnet—the woolly topknot on an alpaca's head and between its ears; it's considered a desirable aesthetic quality; also known as a wool cap.

Break—a spot where fiber is abnormally weaker along its length.

Breed—a race of animals within a species.

Breed back—a breeding with a herd sire from the ranch from which a pregnant female is purchased; offered in contracts as a three-in-one deal.

Britch—the area composed of the lower thighs on a lama's rear legs.

Browse—cellulose-rich food such as leaves, twigs, and weeds; also the act of feeding upon browse.

Cama—a crossbred animal produced by artificially inseminating a female llama using semen from a dromedary camel.

Camelid—members of the camel family including the Old World camelids (camels) and New World camelids (alpacas, llamas, guanacos, vicuñas, and hybrids thereof).

Carding—the final cleaning process before fleece is spun.

Cards—wooden paddles set with short wire pins that are used to card fleece.

Catch pen—a small, well-fenced area used for catching and training lamas.

Ccara (CAR-ah)—the short-wooled "classic" llama; in some places ccara refers to a working llama as opposed to a fiber llama.

Ch'aku (ch'ah-koo)—South American llama term meaning "woolly."

Charki (char-key)—llama jerky; the English word *jerky* is derived from *charki*.

Choanal atresia—a birth defect characterized by the inability to nurse and breathe at the same time; it is caused by an obstruction between the nasal cavity and the throat.

Clean fleece weight—the weight of a fleece after all vegetable matter has been removed.

Clip—the total amount of fiber harvested by a producer in one growing period.

Colostrum—the first thick "milk" produced by a female mammal.

Concentrates—low-fiber, energy-dense feeds such as grain.

Congenital defect—a defect an individual is born with, rather than one it has acquired.

Cotted fiber—matted fiber.

Cover—the act of a female lama being bred by a male.

Cria (CREE-uh)—a young lama between birth and weaning age.

Crimp—waviness along the length of an individual fiber or lock of fleece.

Cud—a bolus of semidigested food brought up from the stomach for rechewing.

Cull—to remove less-desirable animals from a group.

Curaca (cur-AH-cah)—a longer-wooled llama but still within the ccara type.

Cush—an alternate spelling of *kush*.

Dam—an animal's female parent.

Dung piles (often called community dung piles)—the areas where lamas urinate and defecate.

Dust bath (or dust pile)—a bare area on the earth where lamas roll.

Dystocia (dis-TOH-shuh)—difficulty in giving birth.

Felting—intentionally pressing together fibers to form a nonwoven fabric; this can unintentionally occur on the lama as well, resulting in thick mats.

Fiber—lama "wool."

Fiber grades—a system for grading alpaca fiber:

- **royal:** fewer than 20 microns (usually reserved for fashion designers)
- **baby ("premium alpaca"):** 20–23 microns
- **standard:** 24–28 microns
- **coarse:** 29–33 microns
- **inferior:** 33–35 microns
- **very coarse:** more than 35 microns

Fighting teeth—six very sharp canine-like teeth, two in the lower jaw and one in the upper jaw on both sides of a male lama's mouth.

Fineness—a measure (in microns) of the diameter of individual fibers.

Finish—the very end of a lock or a curl.

Fleece weight—the weight of all usable fiber removed from a single animal.

Forage—high-fiber, less energy-dense feeds such as grass and hay.

Gait—how an animal moves. Lamas' gaits are the walk, pace, trot, and gallop.

Gallop—a swift, three-beat gait; it's the fastest of the four lama gaits.

Genetic marker—a detectable gene or DNA fragment.

Genotype—an individual's genetic makeup.

Get—a male animal's offspring.

Get of sire—a show class in which three lamas with the same sire and at least two different dams are shown as a group.

Going down—the act of a female lama dropping into the kushed position prior to being bred.

Guanaco (gh'wah-NAH-coh)—a wild South American camelid; ancestor of the llama.

Guard hair—coarse, medullated (hollow) hair composing a second, outer coat of fiber on llamas (but not alpacas).

Hand—a subjective tactile measure of the softness of fiber.

Handle—*see* hand.

Hembra (h'EM-brah)—a female alpaca.

Herd sire—a male lama used for breeding purposes.

Huacaya (h'wha-k'EYE-ya)—a type of alpaca with crimped, plush fleece.

Huarizo (h'whar-EE-soh)—an alpaca-llama hybrid. Some sources say only a male alpaca bred to a female llama produces a huarizo and the reverse mating of a male llama to a female alpaca produces a misti. Unlike hybrids of other species, these crosses produce fertile offspring.

Hybrid—an individual whose parents are of two different species.

Humming—the droning sound lamas make under a variety of circumstances.

IgG (Immunoglobulin G)—antibodies in the colostrum of near-term female lamas and those that have just given birth.

Induced ovulator—a female animal (including all lamas) that ovulates after, instead of before, being bred.

Kemp—coarse, medullated hair fibers scattered throughout a fleece.

Kush (also spelled *cush*)—the act of a camelid lying down sternally with its legs tucked under it. It is also the name of the position as well as the command given to an animal to signal it to kush.

Lama—a term used to encompass all of the South American camelids, including alpacas, llamas, guanacos, vicuñas, and hybrids thereof.

Lanuda (lah-NOO-dah)—a long-wooled llama with fringes on the ears and abundant wool down the legs.

Llama-guanaco (LAH-ma gh'wah-NAH-coh; or in South America, YAH-ma gh'wah-NAH-coh)—the fertile, crossbred offspring of a male guanaco and a female llama.

Llamo-vicuña (LAH-mo vee-COON-yah; or in South America, YAH-mo vee-COON-yah)—the fertile, crossbred offspring of a male vicuña and a female llama.

Line—a group of related individuals.

Luster (also spelled *lustre*)—the natural sheen of certain fibers.

Macho (MAH-choh)—a male alpaca used for breeding purposes.

Maiden—a female that has never been bred.

Mean—the average fiber diameter of all the fiber in a fleece, measured in microns.

Medulla—the hollow core inside guard hair or kemp.

Medullation—the degree to which a fleece contains medullated fiber.

Micron—a measurement of fiber diameter, equal to 1/25,000 of an inch or 1/1000 of a millimeter. Used to refer to the fineness of a fiber: a smaller micron equates finer fiber.

Midside—a point midway between the front and back legs and slightly lower than halfway down the side of an animal.

Misti (MEES-tee)—the crossbred, fertile offspring of a male llama and a female alpaca.

Neck—fleece similar to blanket but shorter in length.

Nonbreeders—a show term used to designate geldings, as well as females that are certified as nonreproductive.

Open—not pregnant.

Overconditioned—fat.

Pace—a medium-speed, two-beat gait in which both limbs on the same side move in unison.

Packer—a strong, well-conditioned pack llama used for carrying heavy loads for relatively long distances.

Paco-guanaco (PAH-coh gh'wah-NAH-coh)—the fertile, crossbred offspring of a male alpaca and a female guanaco.

Paco-vicuña (PAH-coh vee-COON-yah)—the fertile, crossbred offspring of an alpaca and a vicuña.

Paqocha (pah-KOH-chah)—Quechua for *alpaca*.

Pastern—the lowermost joint in the legs.

Pasture breeding—the breeding system by which a male and a group of females cohabitate in a pasture, allowing mating to occur when it will.

Pedigree—a genealogy of an individual's ancestors.

Pen breeding—the breeding system by which one male and one female are released in a small enclosure for mating purposes.

Polydactylism—having more than the usual number of toes.

Prepotency—the ability of an individual to sire or produce uniform offspring.

Prime fleece—high-quality fleece taken from the part of a lama where a horse blanket would fit.

Produce—a female animal's offspring.

Produce of dam—a show class in which two lamas with the same dam and two different sires are shown as a pair.

Pronk—a stiff-legged, bouncing gait in which all four feet hit the ground at once; lamas pronk while playing.

Proven—an animal that has successfully sired or produced live offspring.

Pureblood—*see* purebred.

Purebred—an individual whose ancestors are of the same breed for a predetermined number of generations.

Q'ara (CAR-ah)—South American llama term meaning "bare"; *see* ccara.

Quechua (KAY-chu-ah)—aboriginal people of Peru, Bolivia, Argentina, Ecuador, and Colombia who speak the Quechuan language.

Roving—a narrow, snakelike, twisted roll of fiber produced during processing before it is spun into yarn.

Saddle—*see* blanket.

Second cuts—short, prickly fibers created when the shearing head runs over the same area of an animal more than once during a single shearing.

Separation—the difference between guard hair and fiber.

Shear weight—*see* fleece weight.

Sire—an animal's male parent.

Skirt—to remove unusable edges from a shorn fleece.

Spin—to twist fiber into yarn; this can be done using commercial machinery, a spinning wheel, or a drop spindle.

Spit test—exposing a previously bred female to a male to see if she will "spit him off" (resist his advances, often by emphatically spitting at him), indicating that she has conceived.

Staple—a group or lock of individual fibers.

Staple length—the length of a group or lock of individual fibers.

Suri (SIR-ee)—a type of llama or alpaca characterized by individual locks of fleece hanging in ringlets.

Syndactylism—having toes that are fused together.

Tags—unusable bits of felted or dirty fleece removed from the lower legs.

Tampada (tam-PAH-dah)—a long-wool llama, but not as woolly as a lanuda.

Tampuli (tam-POO-lee)—a catch term for woolly llamas (tampadas and lanudas).

T'aqa (t'AH-kah)—a llama-alpaca cross that resembles the alpaca parent.

Tensile strength—the amount of force needed to break a fiber staple of given thickness.

Three-in-one package—a pregnant female sold with her unweaned cria.

Tipped ears—ears that are not fully erect.

Topline—the animal's back and croup as viewed from the side.

Trot—a medium-speed, two-beat gait in which diagonal legs move at the same time.

Tui (TOO-ee)—a yearling alpaca; also the first fleece (usually graded as baby alpaca) produced by a young alpaca.

Two-in-one package—an open female sold with her unweaned cria.

Underconditioned—thin.

Unpacking (slang)—giving birth.

Vegetable matter (VM)—Sticks, burrs, hay chaff, and such in an uncleaned fleece.

Vicuña (vee-COON-yah)—a wild South American camelid; ancestor of the alpaca.

Warilla (whar-EE-yah) or wari (WHAR-ee)—a llama-alpaca cross that resembles the llama parent.

Wastiness—the loss of fiber in carding and combing due to vegetable matter, weakness or tenderness, or shortness of fiber.

Weanling—a young lama who has been weaned but is still under one year of age.

Withers—the part of the lama where its back and shoulders meet the neck.

Wool break—a place where fiber breaks easily when subjected to pressure.

Woolen yarn—compared to worsted yarn, woolen yarn is shorter, loftier, and softer. It's usually carded rather than combed, resulting in fibers that go in different directions rather than parallel. This incorporates more air into the yarn. Woolen yarns are used for knitting and weaving fluffy blankets.

Woolies—heavy-wool llamas.

Worsted yarn—worsted yarns are smooth, quite strong, and longwearing, typically spun from fibers that are three inches in length or longer. They are used for woven clothing such as men's suits. The fiber is usually carded, combed with wool combs, and then drawn. This process removes the shorter and keeps the remaining fibers in parallel order.

Yearling—a young lama between one and two years of age.

Resources

There are great resources out there to help you learn more about llamas and alpacas. The following are some of my favorite ones. To compile an up-to-date, nearly free library of llama or alpaca materials, I suggest downloading favorite PDF files to save for future reference. Print bulletins and file them, or bind printouts to create your own "everything about lamas" reference books.

Periodicals

OL = online
PR = print

Llamas

The Backcountry Llama (PR)
http://www.bcllama.com
406-826-2201
The Backcountry Llama is a magazine of llama packing.

Llama Banner (PR)
http://www.llamabanner.com
785-537-0320
Subscriptions to *Llama Banner* include seven great issues and a llama calendar.

Llama Life II (PR)
http://www.llamalife.com
434-286-4494
A tabloid-sized, black-and-white glossy quarterly, *Llama Life II* has been published for more than 20 years.

Alpacas

Show & Tell (PR)
http://www.alpacashowandtell.com
724-863-0909
Show & Tell calls itself "The alpaca exhibitor's magazine showcase."

Alpaca Journal (OL)
http://www.alpaca-journal.com
503-628-3110

Alpaca Journal is Mike Safley's excellent online alpaca magazine; don't miss the hundreds of archived articles at this site.

Alpaca World (PR)
http://www.alpacaworldmagazine.com
01-884-243-579
Alpaca World is a high-quality, independent quarterly based in Great Britain; visit to view hundreds of articles archived online.

Alpacas Magazine (PR)
http://www.alpacaowners.com
615-834-4195
Alpacas Magazine is the official journal of the Alpaca Owners and Breeders Association; it's outstanding!

Llamas and Alpacas

Cool Camelids (PR)
http://www.coolcamelids.org
205-369-3111
Cool Camelids is a high-quality quarterly for owners and admirers of all camelid species.

Hummer Country Webletter (OL)
http://www.hummercountry.org
Hummer Country, a monthly online "Webletter," features excellent monthly articles as well as coming events listings for the United States, Great Britain, Australia, and New Zealand. A must!

International Camelid Quarterly (PR)
http://www.llamas-alpacas.com
866-861-6248
Subscribe to *CQ* and gain online access to hundreds of archived articles in PDF.

Lamalink.com (PR & OL)
http://www.lamalink.com
406-755-5473
Lamalink.com is an online (free) and print (by subscription) llama and alpaca magazine; many issues are on its Web site as free, downloadable PDF files.

Books and CDs/DVDs

Llama Books

Birutta, Gail. *Storey's Guide to Raising Llamas: Care/Showing/Breeding/Packing/Profiting.* Storey Publishing, 1997.
Some of the information in this comprehensive book is dated, but it's still a recommended read. Especially useful: a chapter on establishing a llama-packing service and one on marketing llama manure to gardeners.

Burt, Sandi. *Llamas: An Introduction to Care, Training, and Handling.* Alpine Publications, 1991.
This lavishly illustrated, 208-page book is my favorite all-around llama guide. It's out of print but you can purchase used copies at eBay and Amazon.com for a pittance. A modicum of information is outdated, but author Sandi Burt's publication is still a best buy for novice owners.

Daugherty, Stanlynn. *Packing with Llamas.* 4th ed. Juniper Ridge Press, 1999.
Packing with Llamas is packed with information on selecting, training, and

caring for llamas at home and on the trail. It's the best in-print llama-packing book on the market.

Harmon, David, and Amy S. Rubin. *Llamas on the Trail: A Packer's Guide.* Mountain Press, 1992.
Llamas on the Trail covers everything beginners need to know to start llama packing. It, too, is out of print, but used copies are readily available from sources such as eBay and Amazon.com.

Llama CDs/DVDs

Goldsmith, Bobra. *Llama Training with Bobra Goldsmith: What Every Llama Should Know.* DVD, 115 minutes.
The title is self-explanatory. Bobra Goldsmith shows viewers how to correctly halter an inexperienced or fearful llama and how to train llamas of all ages to lead (and jump!).

———. *Teaching Llamas to Drive.* DVD, 117 minutes.
Bobra Goldsmith demonstrates everything from selecting a driving llama and its equipment, to harnessing and hitching, to teaching the llama to drive.

McGee Bennett, Marty, and Stanlynn Daugherty. *Teaching Your Llama to Pack.* DVD, 75 minutes.
Marty McGee Bennett of CAMELIDynamics teams with veteran packer Stanlynn Daugherty *(Packing with Llamas)* to show how to use TTEAM (Tellington Touch Every Animal Method) techniques to train pack llamas. A bonus: the final section teaches viewers how to trim toenails, shear, and use bodywork techniques on llamas of all kinds.

Spalding, Cathy. *Llama Talk: Understanding Llama Behavior as a Foundation for Training and Herd Management.* CD-ROM.
This fantastic CD textbook by clinician Cathy Spaulding features photos of llama behavior, recordings of llama vocalizations, and a lavish amount of material about llama psychology. Highly recommended!

Alpaca Books

Evans, C. Norman, DVM. *Alpaca Field Manual.* Able, 2003.
Although currently out of print, this extremely comprehensive manual is considered by many alpaca breeders to be the best all-around guide to alpaca care and management.

Hoffman, Eric, et al. *The Complete Alpaca Book.* Bonnie Doon Press, 2006.
More than 600 pages, 1,350 references, 250 tables and figures, 500 photographs, and 3 color sections written by dozens of experts in their field make this a top-flight reference work for alpaca owners. It's expensive—but worth it!

Sheets, Tom, and Beth Sheets. *Alpacas: A Getting Started Guide.* Heritage Farm, 2006.

Tim and Beth Sheets of Heritage Farm Suri Alpacas produced this inexpensive, fifty-one-page, spiral-bound book to help people decide if they want to breed alpacas. It covers everything from alpacas as pets and fiber animals to raising top-flight breeding stock; the section on evaluating fleece is outstanding. Consider this a best buy for prospective alpaca owners. It's available through eBay or directly from the author at http://www.ourheritagefarm.com (765-566-3077).

Alpaca CDs/DVDs

Chepolis, Ted, and Elaine Chepolis. *The Complete Alpaca Shearing Guide to Better Fleeces & Show Success.* DVD, 2 discs.
Topics include: equipment selection; lying down, standing up, and table shearing methods; fleece skirting techniques; fleece assessment; and marketing. Visit the Chepolises' Pine Lake Alpacas Web site (http://www.alpaca-com.com) to view clips from this popular DVD.

McGee Bennett, Marty. *Alpaca Training and Handling.* DVD/VHS, 85 minutes.
Marty McGee-Bennett demonstrates using TTEAM principles to halter and care for alpacas of all ages and levels of training.

Spalding, Cathy. *Alpaca Talk; Understanding Alpaca Behavior.* CD-ROM.
Alpaca Talk is divided into ten sections and forty-five chapters; each section begins with a behavioral guideline listing cues to normal behaviors and to behaviors that may not be normal. Insightful and descriptive text examines these behaviors in detail, accompanied by 110 related color photos of alpaca behavior. Highly recommended!

Llama and Alpaca Books

Fowler, Murray E. *Medicine and Surgery of South American Camelids: Llama, Alpaca, Vicuña, Guanaco.* 2nd ed. Wiley-Blackwell, 1998.
This 549-page veterinary text covers camelid biology; toxicology; diseases; congenital and hereditary conditions; neonatal development; and nutrition, care, and handling; it's the veterinary reference for lama owners.

Hoffman, Claire, DVM, and Ingrid Asmus. *Caring for Llamas and Alpacas: A Health and Management Guide.* 2nd ed. Rocky Mountain Lama Association, 1996.
This great book provides basic information on the daily care of lamas; how to buy, transport, and restrain them; how to manage their health concerns; and how to deliver their young.

MacQuarrie, Kim, Jorge Flores, and Javier Portus. *Gold of the Andes: The Llamas, Alpacas, Vicuñas and*

Guanacos of South America. 2 vols. O. Patthey and Sons, 1994.
This is a lavishly photo-illustrated, 628-page study of the South American camelid in its native home—it's expensive but absolutely gorgeous!

McGee Bennett, Marty. *The Camelid Companion.* Raccoon Press, 2001.
If you only buy one lama training resource, make it this one. Marty McGee Bennett, queen of TTEAM for camelids, packs an astounding amount of information into the 400 pages of this lavishly illustrated, large-format paperback. It's my favorite camelid book, bar none.

Smith, Bradford B., Karen I. Timm, and Patrick O. Long. *Llama and Alpaca Neonatal Care.* Clay Press, 1996.
Chapters in this well-written manual include anatomy, physiology, stages of labor, dystocias, and the care and problems of newborn crias and their dams. Recommended!

Llama and Alpaca CDs/DVDs

McGee Bennett, Marty. *Camelid Culture.* Audio CD, 53 minutes.
Marty McGee Bennett discusses numerous aspects of lama psychology and how it affects the ways in which camelids and their handlers interact. If you're new to lamas you will love this CD.

———. *Getting Started with TTEAM.* DVD/VHS, 54 minutes.
If you aren't familiar with TTEAM (http://www.ttouch.com) but you own camelids, this is the program for you. I've used TTEAM for more than twenty years on horses, dogs, sheep, goats, and now llamas, and I endorse it highly. This DVD makes learning and applying it a breeze.

———. *Understanding Aggression in Camelids.* Audio CD, 50 minutes.
Marty McGee Bennett discusses aberrant behavior syndrome (which she calls novice handler syndrome) and how to prevent it. If you have male camelids, don't overlook this indispensable resource.

———, and David Anderson, DVM, MS. *Camelid Handling Secrets.* Vol. 1, *Medical Management.* DVD/VHS, 60 minutes.
Marty McGee Bennett and David Anderson, DVM, MS, of Ohio State University, explain how to set up a facility in ways that result in the least amount of stress for you, your veterinarian, and your lamas.

———, and LaRue Johnson, DVM, PhD. *Treating Your Lama Kindly.* DVD/VHS, 64 minutes.
Marty McGee Bennett and camelid veterinarian Dr. LaRue Johnson apply TTEAM techniques to a host of veterinary procedures in this interesting program designed for veterinarians and lama owners alike.

Marketing Books

Winslow, Ellie. *Growing Your Rural Business: From the Inside Out.* Self-published, 2008.

———. *Marketing Farm Products: And How to Thrive Beyond the Sidewalk.* Self-published, 2007.

These how-to books by rural marketing maven Ellie Winslow are available in print and e-book form directly from the author via her Web site (http://beyondthesidewalk.com) or at any of her popular rural marketing seminars. It's the most comprehensive material on livestock marketing that I've seen.

Organizations

Llama Organizations

American Miniature Llama Association (AMLA)

http://www.miniaturellamas.com
406-755-3438
The AMLA promotes and registers llamas three years of age or older that measure no more than 38 inches at the withers.

Argentine Llama Aficionados (ALA)

http://www.argentinellamas.org
Argentine Llama Aficionados is a group of breeders dedicated to promoting beautiful, woolly llamas of Argentine bloodlines.

Llama Association of North America (LANA)

http://www.llamainfo.org
The LANA is a group of llama owners and breeders involved in educating the public in many ways, including group advertising.

Pack Llama Trial Association (PLTA)

http://www.packllama.org
The purpose of the PLTA is to preserve and promote pack llamas by educating the public in the safe and humane use of llamas as packing companions. The group also sanctions numerous pack llama trials throughout the United States each year.

Suri Llama Association and Registry

http://www.surillama.com
877-852-1054
The Suri Llama Association and Registry educates the public about suri llamas and maintains a suri llama registry.

Alpaca Organizations

Alpaca Association New Zealand

http://www.alpaca.org.nz
Alpaca Association New Zealand is a stellar source of information about alpacas. Be sure to click on *Alpaca Health & Disease* to download an array of invaluable paddock cards.

Alpaca Owners and Breeders Association (AOBA)

http://www.alpacaowners.com
615-834-4195
You'll find everything you could possibly want to know about alpacas at the AOBA Web site.

Alpaca Registry, Inc. (ARI)

http://www.alpacaregistry.net
402-437-8484
The ARI is the world's largest alpaca pedigree registry.

Suri Network

http://www.surinetwork.org
970-586-5876
The Suri Network is a cooperative of breeders producing rare suri alpacas.

Llama and Alpaca Organizations

Alpaca Llama Show Association (ALSA)

http://www.alsashow.org
412-761-0211
The ALSA promotes alpacas and llamas through showing; it establishes guidelines for shows, educates judges, and records the achievements of individual lamas.

Canadian Llama and Alpaca Association

http://www.claacanada.com
800-717-5262
Browse the Canadian Llama and Alpaca Association Web site's FAQs and articles about lamas.

International Camelid Institute (ICI)

http://www.icinfo.org
614-403-1016
Founded in 2001 by David E. Anderson, DVM, at the Ohio State University College of Veterinary Medicine, the ICI promotes education, service, and collaboration among researchers, breeders, owners, fiber and textile industry professionals, and animal enthusiasts worldwide. Its Web site is a first-class source of in-depth information about alpacas, camels, guanacos, llamas, and vicuñas.

International Lama Registry (ILR)

http://www.lamaregistry.com
406-755-3438
The ILR is a not-for-profit corporation with the purpose of maintaining an official genealogical registry system and research services for owners of subspecies of the genus *Lama:* llama (*Lama glama*), guanaco (*Lama guanicoe*), vicuña (*Vicugna vicugna*), and crossbreds thereof.

New Hampshire Lama Association (NHLA)

http://www.nhlama.org
The New Hampshire Lama Association Web site is an unusually rich source of informative material. Click on any of the topics listed under *Lama Info* for a comprehensive discussion of that subject, and be sure to click *Selected Articles* to access more than 100 archived *NHLA Newsletter* articles.

Rocky Mountain Llama and Alpaca Association (RMLA)

http://www.rmla.com
This site is a valuable resource for all

lama lovers, not just those living in the Rocky Mountain region. Articles covering a huge range of health topics highlight this great site.

Other Organizations

The Paco-Vicuña Registry

http://www.paco-vicunaregistry.com
970-586-4624
The Paco-Vicuña Registry registers paco-vicuñas (alpaca-vicuña crossbreds) and alpacas that resemble vicuñas.

The Pacuña Registry

http://www.vicunaalpacaregistry.com
603-489-9269
Pacuñas are alpaca-vicuña crosses that are DNA-certified to be at least one-quarter vicuña.

Online Article Collections

Llama Articles

Llama Lifestyle Marketing Association
http://www.llama.org
877-425-5262
The not-to-be-missed Llama Lifestyle Web site is jam-packed with scores of resources for llama owners.

Llama Linda Ranch
http://www.paco-vicunas.com
Click on *Articles*; be sure to visit the paco-vicuña pages while you're there.

Llamapaedia
http://www.llamapaedia.com
Llamapaedia is my favorite all-around online information portal; highly recommended!

LlamaWeb
http://www.llamaweb.com
LlamaWeb is another first-rate, comprehensive repository of llama information.

Shagbark Ridge Llamas
http://www.shagbarkridge.com
The Shagbark Ridge Llamas Web site is a peerless resource for llama lovers; there are more than 100 articles archived under *Vet Corner* alone!

Shangrila Llamas
http://www.shangrilallamas.com
Visit the Shangrila Llamas site to peruse and download some great material on clicker training llamas for show (in obstacle classes), on packing, and on llamas as therapy animals.

Alpaca Articles

Alpaca.com
http://www.alpaca.com
Billed as "the world's premier alpaca source and marketplace," Alpaca.com offers free AlpacaFinder services, an online auction, the Alpaca Shop, and an extensive learning center.

Alpaca Association New Zealand
http://www.alpaca.org.nz
Visit and click on *Alpaca Health & Disease* to download invaluable paddock charts on an array of topics about keeping llamas and alpacas in the pink.

AlpacaNation
http://www.alpacanation.com
AlpacaNation is a comprehensive guide to all things alpaca. Don't miss the alpaca library and AlpacaNation's vast pressroom archives.

AlpacaSeller
http://www.alpacaseller.com
Find that perfect alpaca at AlpacaSeller. While you're there, click on *Alpaca Articles* to access scores and scores of fantastic articles from the archives of *Alpaca World* magazine.

AlpacaStreet
http://www.alpacastreet.com
AlpacaStreet is one of the finest alpaca resources on the Internet. In addition to scads of articles, AlpacaStreet boasts alpaca forums and blogs, classified and display ads, a state-by-state guide to camelid veterinarians, fiber resources, interactive tools such as the AlpacaStreet Alpaculator (use it to evaluate potential purchases), and much, much more.

Gateway Farm Alpacas
http://www.gateway-alpacas.com
The Gateway Farm Alpacas Web site is a fantastic source of alpaca material, including the best shearing information online.

HaSu Ranch
http://www.hasu.biz
A click on *Library* leads Web site visitors to an extensive array of alpaca articles.

Maine Alpaca Association
http://www.mainealpacafarms.com
Click on *Alpaca Resources* to download great articles in PDF; don't miss "Knitting with Handspun Alpaca Yarn."

Northwest Alpacas
http://www.alpacas.com
If it's about alpacas, you'll find it at this fantastic site. Don't miss the interactive planner and calculator features and Northwest Alpacas' quarterly *Win an Alpaca* sign-up.

Southern Iowa Alpacas
http://www.southerniowaalpacas.com
More than two dozen great alpacas magazine articles are downloadable from the Southern Iowa Alpacas Web site. Their alpaca FAQ and the ACCOYO Genetics and Investing in Alpacas features are good ones, too.

Lama Feeds and Supplements

Agway Feeds

http://www.agwayfeed.com
Contact is via their Web site.
Agway manufactures two combination llama-alpaca feeds and distributes them throughout the northeastern states.

Blue Seal Feeds

http://www.blueseal.com
800-367-2730
Blue Seal manufactures separate lines of llama and alpaca feeds, minerals, and supplements formulated by camelid expert Dr. Norm Evans. Their products

are available in the Midwest, the Southwest, the Northeast, and the Mid-Atlantic states.

Buckeye Nutrition

http://www.buckeyenutrition.com
800-898-9467
Buckeye Nutrition manufactures Buckeye Llama and Alpaca Concentrate Crumbles and distributes them throughout most eastern and central states.

Chaffhaye

http://www.chaffhaye.com
915-964-2406
Chaffhaye is America's primary haylage product, and it is marketed as "pasture-in-a-bag."

Custom Milling

http://www.custommilling.com
877-348-3048
Georgia-based Custom Milling manufactures Golden Blend llama and alpaca supplements and minerals, as well as Golden Blend Herbal Milk Enhancer for pregnant and lactating lamas.

Dynamite Specialty Products

http://www.dynamitemarketing.com
800-697-7434
Dynamite manufactures alpaca feed; it's available through distributors in most parts of the United States.

Llama Doc Herbs

http://www.llamadocherbs.com
Llama Doc Herbs compounds and sells a variety of herbal supplements for lamas.

Mazuri Exotic Animal Feeds

http://www.mazuri.com
800-227-8941
Mazuri manufactures a full line of feeds, minerals, and supplements for llamas and alpacas. Mazuri products are sold throughout the United States.

Stillwater Minerals

http://www.stillwaterminerals.com
800-255-0357
Stillwater Minerals markets four highly regarded mineral formulations specifically for llamas and alpacas.

United States Department of Agriculture Farm Service Agency's Hay Net Web site

http://www.fsa.usda.gov/FSA/webapp?area=online&subject=landing&topic=hay

U.S. Alfalfa

http://www.usalfalfa.net
620-285-7777
U.S. Alfalfa distributes bagged alfalfa throughout the United States; in some places it's bagged and sold as Purina bagged alfalfa. This is what we feed our llamas, sheep, goats, horses, and donkey. Recommended!

Information, Supplies, and Equipment

Housing

Port-a-Hut

http://www.port-a-hut.com
800-882-4884
Port-a-Hut manufactures lama-friendly steel Quonset-style struc-

tures in a wide variety of sizes. We use them and love them.

Lama Gifts and Supplies

Llama Ridge

http://www.llamas-information.com
T-shirts, books, DVDs.

Lock 'n Luster Grooming Products

http://www.llama-show-tack.com
760-349-9999

Mount Lehman Llamas

http://www.mountlehmanllamas.com
Click on *Lama Trivia* at the Mount Lehman Llamas Web site to access scores of pages of pictures of vintage photos, postcards, stereo views, and trading cards; ethnic jewelry and musical instruments; stamps and currency; pottery; cartoons and much, much more—all featuring llamas!

Nose-n-Toes Llama Gifts

https://secure.nose-n-toes.com
800-530-6391
Look for lama-themed goodies for yourslf or as gifts. These sellers have it all!

Quality Llama Products

https://secure.llamaproducts.com
800-638-4689
Free print catalog.

Ridge Mist Llama and Alpaca Supplies

http://www.llama-alpacasupplies.com
800-24LAMAS
Free print catalog.

Rocky Mountain Llamas Equipment Catalog

http://www.rockymtllamas.com
303-530-5575
Looking for llama driving equipment? It's here! This company carries a huge array of llama and alpaca tack, supplies, and gifts.

Schreiner's Herbal Solution

http://www.schreiners.com
800-223-HEAL
It's our favorite wound dressing.

Stevens Llama Tique

http://stevenstique.com
800-4MY-LAMA (800-469-5262)
Free print catalog.

Useful Llama and Alpaca Items

http://www.lamashop.com
800-635-5262

Packing

Bonny Doon Llama Packs

http://www.llamapacks.com
Bonnie Doon manufactures a trio of excellent packing systems; be sure to read *Tips for the Trail* on the company's Web site.

Buckhorn Llama Company

http://www.llamapack.com
970-667-7411
Buckhorn Llama Company offers llama pack trips and day hikes, sells and leases pack llamas, and markets pack llama gear.

Llama Connection

http://www.packllama2000.com

Wes Holmquist builds llama packing and llama driving gear; click on *Wes' Articles* to read his take on these subjects.

Lost Creek Llamas

home.att.net/~lostcreekllamas/LCL.html

It's a long URL but well worth typing into your Internet browser. The Lost Creek Llamas site is a first-class source of llama information, including specialized material about llama packing and driving.

Mt. Sopris Llamas Unltd.

http://www.soprisllamas.com

Mt. Sopris offers a full selection of packing gear including the beautiful northern white ash, leather, and brass Sopris pack saddle.

Spring Creek Llama Ranch

http://www.springcreekllamas.com

The Spring Creek Llama Ranch site features a dozen useful articles about llama packing.

Driving

Dawn to Dusk Llamas

http://www.dawntoduskllamas.com

Dawn to Dusk Llamas calls itself "Your llama driving resource." Visit the site to buy equipment or llamas, peruse their driving FAQs, or download an excellent llama-driving article in PDF.

Llama Hardware

http://www.llamahardware.com

509-238-2998

Llama Hardware markets "Gear for the working llama," including Eagle carts and Logan driving halters. They carry Flaming Star llama packing gear, too.

Llamas by the Llakes

http://www.llamasbythellakes.com

Visit the Llamas by the Llakes Web site to buy carts, harnesses, and related driving gear; investigate the interesting driving articles while you're there.

Rocky Mountain Llamas Equipment Catalog

http://www.rockymtllamas.com

303-530-5575

Driving clinician Bobra Goldsmith's site offers a full range of llama driving and packing gear, along with an array of everyday essentials and gift items. Be sure to click on *Articles for Llama Lovers* when you visit.

Livestock Guardians

Guard Llamas: A Part of Integrated Sheep Protection

http://www.extension.iastate.edu/Publications/PM1527.pdf

Learn all about llamas as a guardian species in this interesting paper.

Livestock Guardian Dogs

http://www.lgd.org

Here is everything you wanted to know about using livestock guardian dogs to protect your alpacas—and more.

Trainers and Rescues

Lama Training Clinicians

CAMELIDynamics

http://www.camelidynamics.com

Marty McGee Bennett's CAMELIDynamics grew out of her longtime work as a TTEAM (Tellington Touch Every Animal Method) certified instructor. Visit to peruse archived articles; subscribe to the free CAMELIDynamics mailing list; or buy books and videos, training gear, and promotional items from Marty's onsite store. Highly recommended!

Gentle Spirit Behavior & Training for Alpacas and Llamas

http://www.gentlespiritllamas.com

Scope out clinician Cathy Spalding's Gentle Spirit Training site to learn about her wonderful training philosophy and be sure to click on *Articles* while you're there.

John Mallon—Gentling and Training Llamas and Alpacas

http://www.mallonmethod.com

As Marty McGee Bennett's CAMELIDynamics is to Linda Tellington-Jones's groundbreaking work with horses, so John Mallon's prey/predator philosophy is to methods espoused by equine clinician Pat Parelli. Visit his site to read articles and explore the Mallon Method.

Lama Rescues

Brother Francis Llama Rescue and Retirement (WA)

http://www.brotherfrancisrr.org

Colorado Llama Rescue (CO)

303-684-6443

Hog Heaven Rescue Farm (PA)

http://www.hogheavenrescue.org

Adopts out donated alpaca geldings.

Indian Creek Llama Sanctuary (TN)

564 Mahoney Road

Oliver Springs, TN 37840

865-435-4273

Llama Rescue Net (OR)

http://www.llamarescue.org

Northeast Llama Rescue (NY)

http://www.jabed.com/llama

518-827-7733

Southeast Llama Rescue

(active in many states)

http://www.southeastllamarescue.org

Southwest Llama Rescue (NM)

http://www.southwestllamarescue.org

505-690-2611

Check out the great lama training and management material at this Web site!

StillPointe Sanctuary (WA)

http://www.stillpointesanctuary.org

360-452-3656

Western New York Llama Rescue Group (NY)

http://wnyllama.org

360-452-3656

Index

T

U

V

W

X

About the Author

Sue Weaver is author of *Sheep: Small-Scale Sheep Keeping for Pleasure and Profit* and *Chickens: Tending a Small-Scale Flock for Pleasure and Profit*. She also has written hundreds of articles about animals over the years and is a contributing editor of *Hobby Farms* magazine. Sue and her husband, John, live near Mammoth Spring, Arkansas, where as the proprietors of Wolf Moon Boers they raise show-quality full-blood Boer goats. They also raise double-registered miniature American Brecknock Hill Cheviot and Keyrrey-Shee sheep, AMHR Miniature Horses of cob type, and American Curly horses.